Sustaining Urban Networks

The Social Diffusion of Large Technical Systems

Telecommunications, transportation, energy, and water supply networks have gained crucial importance in the functioning of modern social systems over the past 100 to 150 years. *Sustaining Urban Networks* studies the development of these networks and the economic, social, and environmental issues associated with it.

Previous research on industrialized countries has shown that, although many infrastructure networks have become quasi-universal, their development did not spontaneously emerge as a result of technical and economic superiority. Rather the development of networks is the result of complex and often contested dynamics involving systems, uses and users, institutions and territories. The authors analyze challenges to the expansion of access to and use of network-supplied services, as well as challenges associated with such expansion. Far from arguing that expansion is always positive, some of the authors argue that universal development of some networks may prove to be unsustainable.

Analyzing the relations between cities and networks is crucial to discussions of the sustainability of networks and of cities. On the one hand, cities have been, and are increasingly dependent upon the smooth functioning of a host of technological networks; on the other hand, cities are where techological, economic, and social innovations originate, that support the initial development of networks. The functional dependence of cities on infrastructure systems, the social dynamics associated with the initial expansion of a new network in a city, and issues of social/spatial access to basic utility services are analyzed in the chapters of this book.

Sustaining Urban Networks will be of interest to the growing interdisciplinary academic community interested in technological networks, their historical development, their social significance, their role in the functioning of cities, their economic regulation and their expansion in developing countries. It will also be useful reading for strategists in utility companies and governmental agencies.

Olivier Coutard is a Senior Researcher with the French National Center for Scientific Research (CNRS) at Laboratoire Techniques Territoires Sociétés (LATTS), Marne-la-Vallée, France.

Richard E. Hanley is Editor of the *Journal of Urban Technology* and Professor of English at New York City College of Technology of the City University of New York (CUNY).

Rae Zimmerman is Professor of Planning and Public Administration and Director of the National Science Foundation funded Institute for Civil Infrastructure Systems (ICIS) at New York University's Robert F. Wagner Graduate School of Public Service.

The Networked Cities Series

Series Editors:
Richard E. Hanley
New York City College of Technology, City University of New York, US
Steve Graham
Department of Geography, Durham University, UK
Simon Marvin
SURF, Salford University, UK

From the earliest times, people settling in cities devised clever ways of moving things: the materials they needed to build shelters, the water and food they needed to survive, the tools they needed for their work, the armaments they needed for their protection – and ultimately, themselves. Twenty-first century urbanites are still moving things about, but now they employ networks to facilitate that movement – and the things they now move include electricity, capital, sounds, and images.

The Networked Cities Series has as its focus these local, global, physical, and virtual urban networks of movement. It is designed to offer scholars, practitioners, and decision-makers studies on the ways cities, technologies, and multiple forms of urban movement intersect and create the contemporary urban environment.

Moving People, Goods and Information in the 21st Century
The Cutting-Edge Infrastructures of Networked Cities
Edited by Richard E. Hanley

Digital Infrastructures
Enabling Civil and Environmental Systems Through Information Technology
Edited by Rae Zimmerman and Thomas Horan

Sustaining Urban Networks
The Social Diffusion of Large Technical Systems
Edited by Olivier Coutard, Richard E. Hanley, and Rae Zimmerman

Sustaining Urban Networks

The Social Diffusion of Large Technical Systems

Edited by
Olivier Coutard, Richard E. Hanley, and Rae Zimmerman

LONDON AND NEW YORK

First published 2005 by Routledge
2 Park Square, Milton Park, Abingdon, Oxon OX14 4RN

Simultaneously published in the USA and Canada
by Routledge
270 Madison Ave, New York, NY 10016

Routledge is an imprint of the Taylor & Francis Group

Typeset in Times New Roman by
Florence Production Ltd, Stoodleigh, Devon
Printed and bound in Great Britain by
Biddles Ltd, King's Lynn, Norfolk

British Library Cataloguing in Publication Data
A catalogue record for this book is available from the British Library

Library of Congress Cataloging in Publication Data
Sustaining urban networks: the social diffusion of large technical systems/
Edited by Olivier Coutard, Richard E. Hanley and Rae Zimmerman.
p. cm. – (The networked cities series)
Papers presented at a roundtable conference held in New York City in 2001.
Includes bibliographical references and index.
1. Infrastructure (Economics) – Social aspects. 2. Technological innovations –
Social aspects. 3. Public utilities. 4. Municipal services. I. Coutard, Olivier, 1965–
II. Hanley, Richard (Richard E.) III. Zimmerman, Rae. IV. Series.
HC79.C3S95 2004
307.76–dc22 2004008078

ISBN 0–415–32458–0 (hbk)
ISBN 0–415–32459–9 (pbk)

Contents

Notes on Contributors

Bernard Barraqué was trained as a civil engineer. He holds a degree in city planning (Harvard University) and a PhD in urban socio-economic analysis. He is now a full-time researcher in an interdisciplinary research group. His major theme of research covers water-related public policies in Europe, at various territorial levels. He has published the first systematic analysis of water policies in each of the fifteen member states of the European Union (*Les politiques de l'eau en Europe*, Paris: La Découverte, 1995; translated into Portuguese and Italian), as well as several articles in the major European languages. He has also done socio-historical research on various environmental policies. He serves on scientific boards of three French basin authorities, and on several editorial boards in France and in the US (e.g. *Water Policy*, of the World Water Council). He was recently elected as the president of the French national committee of the UNESCO International Hydrological Program.

Olivier Coutard holds an engineer's degree (1988) and a PhD in economics (1994) from the École nationale des ponts et chaussées (Paris). He is a permanent researcher with the French National Center for Scientific Research (CNRS). His research focuses on the social and spatial (especially urban) issues associated with reforms in utility industries (energy and water supply, telecommunications), and on public policies addressing low-income groups' mobility and travel needs. His published books include *The Governance of Large Technical Systems* (ed., London: Routledge, 1999) and *Le Bricolage Organisationnel* (Paris: Elsevier, 2001). He sits on the editorial boards of *Flux* and the *Journal of Urban Technology*.

Ana María Fernández-Maldonado has worked as an architect, urban planner, and researcher in Lima, Peru, and since 1992 in the Netherlands. She holds a position as urban researcher at the Spatial Planning Group of the Section of Urbanism, at the Faculty of Architecture of Delft University of Technology. Since 1997 she has been studying the relationship between information and communication technologies (ICTs) and cities, with special emphasis on large cities of the developing world. She has published both for the academic world and for the development community, contributing among others to *Connected for Development – Information Kiosks and Sustainability*, edited by A. Badshah, S. Khan, and M. Garrido (UN ICT Task Force, 2003) and *The Cybercities Reader*, edited by Stephen Graham (London: Routledge, 2003).

Patrice Flichy is Professor in the Department of Sociology at the University of Marne la Vallée in France, and editor of *Réseaux*, a French journal on communication studies. His research and writing focus on innovation and uses in ICT, both in the past and the present. His published books include *L'Imaginaire d'Internet* (Paris: La Découverte, 2001); *Dynamics of Modern Communication: The Shaping and Impact of New Communication Technologies* (London: Sage, 1995); and *European Telematics: The Emerging Economy of Words* (co-edited with Paul Beaud and Josiane Jouët, Amsterdam: Elsevier, 1991). Patrice Flichy received a PhD in sociology (1971) from the University of Paris 1 Sorbonne. He also holds a degree in business management from the École des Hautes Études Commerciales. He was formerly the head of the sociology research group of France Telecom's Research and Development division.

Stephen Graham is Professor of Human Geography at Durham University in the UK. In 2002/3 he was Deputy Director of the Global Urban Research Unit (GURU) at Newcastle University. He has a degree in geography, an MPhil in planning and a PhD in science and technology policy. His research develops a critical and "socio-technical" perspective to the reconfiguration of urban infrastructures and mobility systems; the growth of urban surveillance; the technological dimensions of social exclusion; the development of urban technology strategies; the "relational" turn in urban theory; and the interconnections between war, terrorism, and cities. As well as a large collection of articles and reports, he is the author of *Telecommunications and the City* (London: Routledge, 1996), *Splintering Urbanism* (London: Routledge, 2001) (both with Simon Marvin), editor of the *Cybercities Reader* (London: Routledge, 2003), and co-editor of *Managing Cities: The New Urban Context* (London: John Wiley, 1995) and *Cities, War and Terrorism* (Oxford: Blackwell, 2003). Between 1999–2000 he was Visiting Professor at MIT's Department of Urban Studies and Planning. Currently he sits on the editorial boards of the *International Journal of Urban and Regional Research, Information, Communication and Society, the Journal of Urban Technology,* and *Surveillance and Society.*

Simon Guy's research interests revolve around the social production and consumption of technology and the material environment. He has undertaken research into a wide spectrum of urban design and urban technology issues including the development of greener buildings, the role of architecture and property development in urban regeneration and the links between building design, mobility planning and the provision of infrastructure services funded by the UK's Economic and Social Research Council and Engineering and Physical Research Council, and the European Union. His publications include *Development and Developers: Perspectives on Property* (ed., with J. Henneberry, Oxford: Blackwell, 2002); *Urban Infrastructure in Transition: Networks, Buildings, Plans* (with S. Marvin and T. Moss, London: Earthscan, 2001), and *A Sociology of Energy, Buildings and the Environment: Constructing Knowledge, Designing Practice* (with E. Shove, London: Routledge, 2000).

Richard E. Hanley is the founding editor of the *Journal of Urban Technology* and the editor of *Moving People, Goods, and Information in the Twenty-First Century*, the first book in the Networked Cities series. He is Professor of English at New York City College of Technology of the City University of New York.

Marie Llorente is working as an independent consultant in environmental economics. She received her PhD in 2002 from the University Paris X. Her dissertation is a contribution to the understanding of the urban water sector crisis in Delhi, based on field-work and interviews. Her current research deals with regulatory issues in the water sector (in France and developing countries) and focuses on participatory mechanisms within decisional process.

Dominique Lorrain is a senior researcher with the French National Center for Scientific Research (CNRS) at the Centre d'étude des mouvements sociaux (CEMS-EHESS). His current work addresses reforms in infrastructure industries (liberalization policies, corporate strategies). He recently edited three special issues in academic journals: "Eau: le temps d'un bilan," *Flux* 52–53 (July–September 2003); "Gouverner les très grandes métropoles" (with Patrick Le Galès), *Revue française d'administration publique*, 107 (December 2003); "Les grands groupes et la ville," *Entreprises et Histoire* 50 (September–November 2002). He is a member of the editorial boards of three academic journals (*Sociologie du Travail*, *Flux*, *Annales de la Recherche Urbaine*) as well as of several scientific committees including the Institut de la Gestion Déléguée and the Technical Advisory Panel of the World Bank's Public–Private Initiative Advisory Facility (PPIAF) program. He teaches the socio-economics of utility industries at the Paris Institut d'Etudes Politiques and in the Tong Ji University (Shanghai) international MBA.

Seymour J. Mandelbaum is Professor of City and Regional Planning at the University of Pennsylvania. For many years he has been principally concerned with planning theory and the design of communities and institutions. He contributed an essay on "Cities and Communication" to the

1988 volume on *Technology and the Rise of the Networked City in Europe and America*. His major work is *Open Moral Communities* (Cambridge, MA: MIT Press, 2000). He is co-editor of a volume on the networked society, scheduled for publication in 2005.

Simon Marvin is co-director of the Centre for Sustainable Urban and Regional Futures (SURF) and United Utilities Chair of Sustainable Urban and Regional Development, both at the University of Salford in the UK. Simon's research has built a detailed understanding of the changing relations between cities, regions, and infrastructure networks in a period of rapid technological change, environmental concern and institutional restructuring. With funding from the two UK research councils he has developed a specialist knowledge of how the reconfiguration of water, waste, energy, telecommunications, and transportation networks re-shapes relations with users and places in contemporary cities and regions. Recent books published include *Urban Infrastructure in Transition: Networks, Buildings and Plans* (with Simon Guy and Tim Moss, London: Earthscan, 2001) and *Splintering Urbanism: Networked Infrastructures, Technological Mobilities and the Urban Condition* (with Stephen Graham, London: Routledge, 2001).

Beth Perry is a Research Fellow at the Centre for Sustainable Urban and Regional Futures (SURF Centre) at the University of Salford. Her main conceptual interests are in urban and regional policy and governance, particularly in relation to theories of multi-level governance and the role of universities in regional development and the knowledge economy. Recent publications include "Universities, Localities and Regional Development: The Emergence of the Mode 2 University?" (*International Journal of Urban and Regional Research*, March 2004), co-authored with Michael Harloe. Her current research is on comparative regional science policies in the context of the European Research Area, funded through the ESRC Science in Society program.

Gene I. Rochlin is Professor in the Energy and Resources Group at the University of California, Berkeley. His research interests include science, technology, and society; cultural and cognitive studies of technical operations; the politics and policy of energy and environmental matters; and the broader cultural, organizational, and social implications and consequences of technology – including large technical systems. His book *Trapped in the Net: The Unanticipated Consequences of Computerization* (Princeton, NJ: Princeton University Press, 1997) won the 1999 Don K. Price Award of the Science, Technology and Environmental Politics Section of the American Political Science Association.

Graciela Schneier-Madanes is an architect (University of Buenos Aires) and a geographer (University of Paris 1 Sorbonne). She holds a permanent position with the French National Center for Scientific Research (CNRS) at the

Research and Documentation Center on Latin America (CREDAL). She is also the director of the CNRS-supported international research network "rés–eau–ville" (water–cities–territories). She teaches at the Institut des Hautes Etudes de l'Amérique Latine (University of Paris 3 Sorbonne Nouvelle) and at the School of Architecture Paris la Villette. Her work centers on urban affairs in several Latin American countries, with research projects related to architectural issues and to network services, in particular the internationalization of urban water management. Her published works include *Eaux et réseaux, les défis de la mondialisation* (ed., with Bernard de Gouvello, Paris: IHEAL/La Documentation Française, 2003); *Buenos Aires, portrait de ville* (Paris: Institut Français d'Architecture/CNRS, 1996); and *L'Amérique Latine et ses télévisions, du mondial au local* (Paris: Economica, 1995).

Sally Wyatt works in the Department of Communication Studies, University of Amsterdam. She is also the President of EASST, the European Association for the Study of Science and Technology. Together with Flis Henwood, Nod Miller, and Peter Senker, she edited *Technology and In/equality: Questioning the Information Society* (London: Routledge, 2000). She also contributed three chapters to *Cyborg Lives? Women's Technobiographies* (eds Flis Henwood, Helen Kennedy, and Nod Miller, York: Raw Nerve, 2001. Her current research is about the role of the internet in the ways in which people construct risks associated with health problems and treatments.

Rae Zimmerman is Professor of Planning and Public Administration at New York University, Robert F. Wagner Graduate School of Public Service. She is Principal Investigator and Director of the NSF-supported Institute for Civil Infrastructure Systems (ICIS). Her teaching and research interests are in environmental planning, management, and impact assessment with a special focus on environmental health risk assessment, environmental equity, and institutional aspects of global climate change; urban infrastructure and its measurement and performance in light of social objectives; and risk management and public perceptions of complex technologies. Most recently she has addressed the security of infrastructure systems in urban areas, and leads NYU's partnership for the Center for Risk and Economic Analysis of Terrorism Events (headquartered at the University of Southern California), funded by the US Department of Homeland Security. Her recent publications include: "Social and Environmental Dimensions of Cutting-Edge Technologies," in *Moving People, Goods and Information in the Twenty-First Century* (edited by R. Hanley, London: Routledge, 2003); "Public Infrastructure Service Flexibility for Response and Recovery in the September 11th, 2001 Attacks at the World Trade Center," in *Beyond September 11th: An Account of Post-Disaster Research* (Boulder: University of Colorado, 2003), and *Digital Infrastructures: Enabling Civil and Environmental Systems Through Information Technology* (ed., with T. Horan, London: Routledge, 2004).

ACKNOWLEDGMENTS

The chapters of this book have been selected from a broader collection of papers presented and discussed at a roundtable conference on the "social sustainability of technological networks," which was held in New York City in April 2001. The conference was organized by the Institute for Civil Infrastructure Systems (ICIS, New York University)* and the Laboratoire Techniques, Territoires, Sociétés (LATTS, Marne-La-Vallée), and co-sponsored by the *Journal of Urban Technology* and *Flux (Cahiers scientifiques internationaux Réseaux et Territoires)*, the leading English- and French-speaking academic journals dedicated to a territorial/urban approach to network infrastructures.

As discussants of the papers presented at the conference, Pierre Bauby (Electricité de France), Michel Gariépy (Université de Montréal), Thomas P. Hughes (University of Pennsylvania), Dominique Lorrain (CNRS), Seymour Mandelbaum (University of Pennsylvania), Jane Summerton (University of Linköping), Ruth Schwartz Cowan (State University of New York at Stony Brook), and Joel A. Tarr (Carnegie Mellon University) have each provided a precious contribution to this intellectual venture. The help of Nate Gilberson (then project manager at ICIS) in the preparation and in the course of the conference was invaluable.

After the conference, Dominique Lorrain, Jane Summerton, and Joel Tarr helped the three editors to select and review the papers that were – often substantially – revised and updated to form the chapters of this book. Neil O'Brien and Wendy Remington greatly contributed to the preparation of the manuscript.

Several public and private institutions in France have provided intellectual, material, and financial help for this project: the Centre National de la Recherche Scientifique (CNRS, Département des Sciences de l'Homme et de la Société); the Centre de prospective et de veille scientifique (CPVS) of the Ministère de l'Equipement, des Transports et du Logement (Direction de la Recherche et des Affaires Scientifiques et Techniques); the École Nationale des Ponts et Chaussées; Electricité de France; and Réseau Ferré de France.

Olivier Coutard, Richard E. Hanley, Rae Zimmerman
Marne-la-Vallée and New York City, March 2004.

Early versions of Chapters 1, 4, 7, 9, and 12 were originally published in the *Journal of Urban Technology*, December 2001, vol. 8, no. 3, Carfax Publishing, an imprint of the Taylor & Francis Group.

Note

* This work is supported by the Institute for Civil Infrastructure Systems (ICIS) at New York University (in partnership with Cornell University, Polytechnic University of New York, and the University of Southern California). This material is based upon activities supported by the National Science Foundation under Cooperative Agreement No. CMS-9728805. Any opinions, findings, and conclusions or recommendations expressed in this document are those of the author(s) and do not necessarily reflect the views of the National Science Foundation.

INTRODUCTION

Network Systems Revisited: The Confounding Nature of Universal Systems

Olivier Coutard, Richard E. Hanley, and Rae Zimmerman

This book is a contribution to the study of the development of the telecommunications, transportation, energy, and water supply, networked systems – sometimes referred to as large technical systems (LTSs) – that have gained crucial importance in the functioning of modern social systems over the past 100 to 150 years.

Previous research on industrialized countries has shown that, although many infrastructure networks have become quasi-universal, their development was not the spontaneous result of their technical and economic superiority. Rather, the development of networks is best understood as the result of a complex process of co-construction of systems, use(r)s and institutions. In line with this tradition, the authors in this book seek to escape deterministic views of the development of infrastructure networks and their "effects" on society. They consider, in particular, that new technologies do not mechanically produce social change, that it is not "in the nature" of LTSs to grow irresistibly, and that network development is a fundamentally contested process. At the same time, they also seek to escape "social determinism," i.e. the idea that the development of technical systems and their role in society are entirely determined by the interplay of "pure" (non-technical) social forces. Rather, the authors in this book would agree to the idea that society is, to a certain extent, determined by technologies in use (Edgerton 1998). They believe that technologies are shaped by society at the same time as they shape society or, in other words, that (social) technical systems and (technical) societies co-evolve.

Building upon the knowledge (both empirical and theoretical) in this area, the authors in this book investigate the development of LTSs in light of sustainability, i.e., they explore the economic, social, and environmental issues associated with the long-term development of those systems (and, often, their universality). They discuss challenges *to* the expansion of access to and use of network-supplied services, as well as challenges *associated with* such expansion, from a sustainability perspective. (Indeed, several authors argue that the universalization of some networks may prove to be unsustainable.)

Many chapters emphasize the urban dimensions of networks. Analyzing the relations between cities and networks is crucial to discussions of the sustainability of networks (and of cities too!). On the one hand, cities have been, since the middle of the nineteenth century, and are increasingly dependent upon the smooth functioning of a host of technological networks; on the other hand, cities are the loci and the foci of technological, economic, and social innovations that sustain the initial development of networks. The book discusses, for example, the functional dependence of cities on networks, the social dynamics associated with the initial expansion of a new network in a city, and issues of social/spatial access to basic utility services. Chapters in the book emphasize the importance of network-shaped and network-shaping uses as well as the importance of institutions in sustaining infrastructure networks.

The study of the sustainable development of LTSs raises a broad range of issues including: the nature and the role of "mediators" between emerging technologies and evolving social behaviors; the conceptions of solidarity or of general interest embedded in or affected by the regulation of network industries and the provision of network services; whether social behaviors, expectations, or values are shaped by networks, and if so how and to what extent; the costs incurred by the dependence of urban and social systems upon networks, and the potential ways to mitigate such costs; the economic, social, and environmental risks associated with the performance, or failure, of networks; the comparative performance of networks and of alternative forms of provision of essential services. The chapters of this book, many based on in-depth empirical studies, explore many of these issues. Despite a common theoretical background, robust areas of contention appear among the authors. Such controversies should be regarded as a resource, rather than an obstacle, in investigating the sustainable development of urban networks.

Networks in Spatial and Urban Systems

The three chapters in Part I discuss the role of network infrastructures in spatial and urban dynamics. This is a controversial area of research. A major contribution to this field is the work of Stephen Graham, Simon Guy, and Simon Marvin, originally associated with the Center for Urban Technologies (CUT) at the University of Newcastle. It was synthesized in a recently published book by Graham and Marvin, *Splintering Urbanism*, in which the authors argue that the "modern integrated infrastructure ideal" is collapsing and with it the drive "to construct ubiquitous, normalized and standardized infrastructure networks," and that "infrastructure networks are [currently] being 'unbundled' in ways that help sustain the fragmentation of the social and material fabric of cities" (Graham and Marvin 2001: 33, 88, 90). The ideas developed in *Splintering Urbanism* run through this book's first three chapters.

In an essay on the history and regulation of networks, Dominique Lorrain analyzes the relationship between networks and cities over time. He argues that we have been entering, over the last two or three decades, a new phase of urban history with the emergence of the gigacity, a new, distinctive form of networked city (Tarr and Dupuy 1988) differing from its nineteenth-century ancestor by its unprecedented size (population), its vertical extension above and below ground, its network density and its blurring of city boundaries made possible by new fast transportation and broadband telecommunications systems. Lorrain relates the advent of this third stage in urban history to the dynamics of network development. Once adequate institutions and rules had been designed by public authorities, he argues, the expansion of "successful" networks was primarily a self-sustained process fueled by the "logic" of utility companies (their strategic interest), scale, and club effects produced by network infrastructures and services, and the development of a very diverse set of network-dependent subsystems, appliances, and social practices. Network services are thus tending to become ubiquitous in contemporary cities, Lorrain argues, refuting the cherry picking (Graham and Marvin 1994) and splintering urbanism theory. In gigacities entirely criss-crossed by infrastructure networks, Lorrain concludes, the major regulatory issues are therefore not about access disparities, but about the reliability of network systems, the contents of network services, and the protection of people's privacy.

In contrast to Lorrain's essay, the chapter by Stephen Graham and Simon Guy emphasizes the exclusionary logic of contemporary network development. Graham and Guy offer a fascinating study of the contested "Internetting" of some of San Francisco's downtown neighborhoods (mainly SOMA and the Mission Area) in the late 1990s. The migration of dot-com entrepreneurs, mainly from the Silicon Valley, to those areas, together with massive investment in telecommunication infrastructures, fueled a major increase in rental values, changes in building uses (with the development of broadband connected "live–work spaces"), as well as "divisive" effects on local communities. The process was highly contested, with fights at the San Francisco planning commission and building occupations in response to threats of eviction. Opposition movements gave rise to attempts at regulating the real-estate boom and its social consequences, and to a broader distrust of development policies within the city's population. The "Internetting" of San Francisco, and the "biased and exclusionary appropriation of selected central urban spaces" that went along with it is, the authors argue, an expression of the more fundamental "shift to a post-national phase of infrastructural development which tends, very broadly, to undermine, or at least challenge, the relatively standardized and equitable infrastructure traps associated with following only the powerful actors (producers) or with accepting too readily the social normalization that is part and parcel of the "imperative to connect."

Part IV of the book contains three contributions to contemporary issues on the development of water supply systems. To a certain extent, they support Wyatt's argument that being connected is not always better. Admittedly, poor access to water is a clear and unambiguous sign of deprivation; in this respect, it cannot be compared to a lack of access to the internet, a more relative form of deprivation. But the three following chapters challenge, to a certain extent, the assumption that physical connection to the network warrants access to the service, and, more fundamentally, the assumption that a networked-based domestic supply of water (and sewerage) should, in principle, be regarded as the universal norm of access to water. In doing so, they emphasize the limits, in analytical as well as in policy terms, of a (rarely thus phrased) notion of "water divide."

At first glance, the two successive water conflicts in Buenos Aires analyzed by Graciela Schneier-Madanes suggest a clear divide, and even opposition, between networked water supply haves and have-nots. Indeed, the first conflict (in 1995) consisted of fierce opposition by populations in network expansion areas to the very high connection charge requested by the water company (in agreement with its concession contract). And the second conflict (in 1998) similarly involved already connected groups who refused to pay for the expansion of the network. It would seem that these conflicts reveal the antagonistic interests of connected and non-connected groups, as well as the deeper social significance of being connected, both in terms of user rights and in terms of social inclusion. However, as Schneier-Madanes shows, those conflicts were rooted in a context in which the population's initial support of privatization reforms (in many utility and public services) was progressively undermined by rate increases for many public services, the lack of subsidies to low-income families, and the new commercial character of the services. This situation was exacerbated by the impoverishment of large parts of the population in the context of a broader economic crisis. The socially inclusive properties associated with the connection to utility networks are hampered by the risk of being disconnected faced by a growing part of the city's population; and being *dis*connected may well be more stigmatizing than *not being* connected in the first place. This corroborates a conclusion of many studies: the physical connection to a centralized or "bundled" network is not by itself the ultimate solution to problems of access to essential services.

In a different institutional context, Marie Llorente discusses the reasons for the poor performance of the formally "bundled," publicly owned and operated water supply system in Delhi, and how this performance might be improved. Delhi's water supply is characterized by an insufficient and low-quality resource, poor condition of infrastructure (with massive wastage), insufficient and intermittent supply, strong disparities in access to and consumption of water, which affect low-income users disproportionately, and have poor cost recovery. Llorente includes in her discussion the many facets of this situation: rules, both "to construct ubiquitous, normalized and standardized infrastructure networks," and that "infrastructure networks are [currently] being 'unbundled' in ways that help sustain the fragmentation of the social and material fabric of cities" (Graham and Marvin 2001: 33, 88, 90). The ideas developed in *Splintering Urbanism* run through this book's first three chapters.

In an essay on the history and regulation of networks, Dominique Lorrain analyzes the relationship between networks and cities over time. He argues that we have been entering, over the last two or three decades, a new phase of urban history with the emergence of the gigacity, a new, distinctive form of networked city (Tarr and Dupuy 1988) differing from its nineteenth-century ancestor by its unprecedented size (population), its vertical extension above and below ground, its network density and its blurring of city boundaries made possible by new fast transportation and broadband telecommunications systems. Lorrain relates the advent of this third stage in urban history to the dynamics of network development. Once adequate institutions and rules had been designed by public authorities, he argues, the expansion of "successful" networks was primarily a self-sustained process fueled by the "logic" of utility companies (their strategic interest), scale, and club effects produced by network infrastructures and services, and the development of a very diverse set of network-dependent subsystems, appliances, and social practices. Network services are thus tending to become ubiquitous in contemporary cities, Lorrain argues, refuting the cherry picking (Graham and Marvin 1994) and splintering urbanism theory. In gigacities entirely criss-crossed by infrastructure networks, Lorrain concludes, the major regulatory issues are therefore not about access disparities, but about the reliability of network systems, the contents of network services, and the protection of people's privacy.

In contrast to Lorrain's essay, the chapter by Stephen Graham and Simon Guy emphasizes the exclusionary logic of contemporary network development. Graham and Guy offer a fascinating study of the contested "Internetting" of some of San Francisco's downtown neighborhoods (mainly SOMA and the Mission Area) in the late 1990s. The migration of dot-com entrepreneurs, mainly from the Silicon Valley, to those areas, together with massive investment in telecommunication infrastructures, fueled a major increase in rental values, changes in building uses (with the development of broadband connected "live–work spaces"), as well as "divisive" effects on local communities. The process was highly contested, with fights at the San Francisco planning commission and building occupations in response to threats of eviction. Opposition movements gave rise to attempts at regulating the real-estate boom and its social consequences, and to a broader distrust of development policies within the city's population. The "Internetting" of San Francisco, and the "biased and exclusionary appropriation of selected central urban spaces" that went along with it is, the authors argue, an expression of the more fundamental "shift to a post-national phase of infrastructural development which tends, very broadly, to undermine, or at least challenge, the relatively standardized and equitable infrastructure systems that were constructed in western nations during the Fordist-Keynesian post war-boom." Although the tensions around the appropriation of space in central San Francisco receded as a consequence of the dot-com failure in the early 2000s, the questions raised by these conflicts remain.

The chapter by Graham and Guy thus offers a carefully documented and reflexive exploration of the splintering urbanism thesis articulated by Graham and Marvin (2001). In the next chapter, Olivier Coutard develops a critique of this thesis. Based on a discussion of historical and contemporary empirical material, Coutard argues, first, that recent reforms in utility industries have not significantly challenged existing universal services in developed countries. Nor have they systematically aggravated the social disparities in access to basic network services in developing countries. More specifically, the notion of "unbundling" used by Graham and Marvin is misleading when applied to network infrastructures in developing countries, insofar as it suggests that the provision of basic services was previously "bundled." In fact, non-network forms of service provision must be included in the picture, as they characterize the everyday life of a majority of the population in those countries. Second, Coutard contends that, contrary to Graham and Marvin's assumption, disparities between spaces in the provision of, access to, and use of network infrastructures are not always socially undesirable. For example, it is not a priori shocking that business districts should benefit from enhanced transportation, telecommunications, and other infrastructure services; the key policy issue is the extent to which the economic achievements of these districts benefit the surrounding population. Third, Coutard contests that infrastructure "unbundling" plays a leading role in residential segregation or in other forms of "privatization" of urban space. Premium network supplies may not even be a good indicator of premium spaces because homogeneous and standardized infrastructures can coexist with strong social or functional specialization of city spaces. Applied to the contested "Internetting" of San Francisco studied by Graham and Guy in the previous chapter, this critique would suggest a close examination of the network specificity, if any, of what is first and foremost a gentrification process.

Risks and Crises in Networked Systems

Part II addresses how risks and crises play out in highly networked systems in urban areas, and how such systems can be sustained (or sustain themselves) in the face of major disruptions. Crises provide opportunities to analyze how (and how much) cities are functionally dependent upon networks. Two chapters analyze examples of network failures and discuss the origins of these failures, the failure processes, their effects, and ways to mitigate their adverse consequences. The two chapters clearly illustrate the risks associated with the ubiquity of networks: functional and physical interdependencies that may lead to systemic, large-scale failures; and deep socially disruptive effects of failures. But they also emphasize the resilience of matrix-pattern networks in dense urban areas and the crucial importance of user responses in the mastery of crises and, at least potentially, in shaping more sustainable future development of infrastructure systems.

The chapter by Rae Zimmerman focuses on interdependencies between infrastructure systems (utilities, roads, computer-based systems). Based on a discussion of various examples, Zimmerman addresses three dimensions in turn: functional and spatial interconnectedness, redundancy in and between infrastructures, and system knowledge aspects. She argues that interconnectedness and interdependencies within and between infrastructure systems are a key element in system performance as well as system vulnerability. This results in tricky technological and managerial issues that are discussed in the chapter along with possible responses to these issues: technical responses such as trenchless technologies that minimize disruptions on road systems caused by utility networks' building or maintenance; regulatory responses such as obligations for utility firms to coordinate their demands on local authorities; responses involving the detailed configuration of computer-based knowledge systems; and organizational and institutional responses (from shared knowledge systems to integrated multi-utility firms).

The next chapter by Simon Marvin and Beth Perry takes a different stance by focusing on the consequences of an infrastructure failure. Based on a study of how a sample of working urbanites dealt with a fuel supply disruption (the British "fuel crisis" of 2000), the chapter examines the implications of the increasing social dependence upon – increasingly vulnerable? – infrastructure systems (in this instance the automobile system). It does so by addressing three issues: How did a sample of car users cope during the fuel crisis and the disruption of both the public and private transport services? What external conditions influenced individual strategies? And to what extent did the alternative travel and behavioral patterns developed during the crisis become embedded in new routines? The study shows, first, that car users were able to develop viable strategies for reducing their motoring, and a range of more sustainable and environmentally friendly transport behaviors emerged; second, that employers facilitated or negated the coping efforts of their employees by providing favorable or non-supportive environments for adjustment; and third, that in that case, the crisis did not last long enough for new behaviors to become embedded. This research corroborates the observations made in a similar situation created by a month-long major strike in the Paris public transport system in 1995: that situation also induced innovative behaviors that did not outlast the end of the crisis. It thus provides useful insights on the plasticity and resilience of individual and social behaviors, a key element in the resilience of networked cities and societies.

A Focus on Two Infrastructure Sectors

Parts III and IV focus on two specific infrastructure sectors: the internet and water. These two sectors differ in several important respects. Technologically, new information and communication technologies (ICTs), including the internet, are characterized by rapid technological innovation and proliferating infrastructures, while drinking water supply and sanitation systems are characterized by stable technologies (not precluding incremental innovations) and well-developed long-lasting infrastructures. And the internet is an emerging and rapidly changing system, whereas water is a relatively old, stabilized service. Functionally, the internet uses the infrastructure of a tier system (making it belong to the group of second-order LTSs analyzed by Braun and Joerges (1994)), whereas water supply mostly uses proprietary infrastructures. Socially, ICTs support a host of economic and social activities, while water is a relatively straightforward service. Economically, ICTs are the sector where competition, local and global, is the most thriving, while water supply is the least liberalized of all network industries. Focusing on sectors at opposite ends of the evolutionary scale of networks has the great advantage of revealing clear-cut, contrasted patterns that would be more blurred in intermediary systems such as energy or urban transportation systems.

Part III of the book thus consists of three chapters on the development of the internet, emphasizing the issues raised by the perspective of the generalization of this already widespread yet relatively new service. Through a careful and penetrating analysis of early discourses on the internet, Patrice Flichy shows the remarkably large extent to which use rules designed by the small community (mostly from academic and counter-culture groups) of so-called *digerati* in the early days of the internet are still valid today in the large communication system the internet has become. Flichy argues that the idealistic social world envisioned by internet pioneers, in which relations between individuals would be equal and cooperative and information would be free, was admittedly challenged as the internet spread. Inequalities in skills (in the use of computing and the production of discourse) of a far greater dimension than in the academic world, have appeared. And the idea of a free internet has faded with the need to finance certain resources through self-supporting mechanisms such as subscriptions. But the initial model has, nevertheless, lasted. Forums for the public at large have been set up, information organized by universities is accessed by different users, and ordinary individuals create sites that present information that is sometimes very valuable. Flichy's analysis, therefore, goes beyond the genealogy of the internet's use rules; it documents and analyzes the emergence of a "network ideology" that is remarkably consistent with widespread social values, expectations and relational patterns.

However, a powerful and widely acknowledged network ideology does not, by itself, warrant widespread access to the corresponding network service. In her examination of the diffusion of the internet in Lima, Peru, Ana María Fernández-Maldonado uncovers some of the mediations that related social values and expectations to the emerging internet system. The chapter first provides a general view of the social diffusion of ICTs (fixed telephone, mobile telephone, cable television, personal computer, and domestic internet access) in the population of Lima, a strongly polarized city. Statistics of domestic or individual access to ICTs reveal a profound "digital polarization" strongly correlated with the city's socio-economic polarization. Fernández-Maldonado then focuses on *cabinas de internet*, the local form of internet cafés. Despite a complete lack of governmental support, *cabinas* developed very quickly, first in higher-income areas of the Peruvian capital, then in the lower-income areas as well, as the result of thousands of individual initiatives from within a predominantly informal local economy. *Cabinas* are very successful, especially among the younger part of the population with a higher-than-average education. Initial motivation for the use of the internet in *cabinas*, which was centered on work, school, and academic purposes, endured, even though users progressively discovered and exploited the communications opportunities offered by the internet. Finally, Fernández-Maldonado critically discusses the significance of ICTs for the improvement of the daily life of poor Lima residents. She notes the many changes associated with the diffusion of *cabinas*: people have been eager and able to improve their ICT-literacy, people go to the *cabina* as their primary recreational activity, and people view the internet as their "window onto the world." *Cabinas* also serve as urban resource centers that are lacking in those areas. Thus, based on their expectations and, in a sense, on their adherence to the "network ideology" described by Flichy, Lima's poorer groups have taken the first step into the "digital economy." But Fernández-Maldonado argues in conclusion that *cabinas* will only allow the achievement of more sustainable goals of local economic development if they benefit from institutional support by local and national government, a support that has, until now, been lacking.

Another form (or dimension) of the "network ideology" is the idea that universal access is unquestionably good. Citizens, policy-makers, and especially researchers should be cautious not to fall into the traps of this preconception. In a very stimulating chapter, Sally Wyatt challenges the widely shared assumption that having internet access is always better than lacking it, and that once financial and ICT-literacy issues have been overcome by cheaper services and education and training, people will embrace the technology wholeheartedly. She does so by symmetrically exploring "the use and non-use" of the internet. She first discusses what she terms "two fallacies" associated with notions of trickling-down or catching-up of internet diffusion. The first is that growth will lead to a more even distribution of users, whereas, Wyatt argues, most of the available data suggest that it does not. Although gender differences in internet access and use have dramatically declined, differences between countries and differences based on race and income remain stark. A second fallacy implicit in the trickle-down assumption about continued growth, Wyatt argues, is precisely that growth will indeed continue. Recent studies provide evidence of a flattening of internet growth in Europe and the US. The possible reasons for this and for the existence of non-users (including voluntary ones) are manifold: high levels of connection costs, the need for a computer, "a potential gap between heightened expectations and the reality of the 'internet experience'," and the declining amount of social prestige that can be gained from being an internet user. In the conclusion, Wyatt highlights the importance of incorporating non-users, together with users, into technology studies as a way of avoiding the traps associated with following only the powerful actors (producers) or with accepting too readily the social normalization that is part and parcel of the "imperative to connect."

Part IV of the book contains three contributions to contemporary issues on the development of water supply systems. To a certain extent, they support Wyatt's argument that being connected is not always better. Admittedly, poor access to water is a clear and unambiguous sign of deprivation; in this respect, it cannot be compared to a lack of access to the internet, a more relative form of deprivation. But the three following chapters challenge, to a certain extent, the assumption that physical connection to the network warrants access to the service, and, more fundamentally, the assumption that a networked-based domestic supply of water (and sewerage) should, in principle, be regarded as the universal norm of access to water. In doing so, they emphasize the limits, in analytical as well as in policy terms, of a (rarely thus phrased) notion of "water divide."

At first glance, the two successive water conflicts in Buenos Aires analyzed by Graciela Schneier-Madanes suggest a clear divide, and even opposition, between networked water supply haves and have-nots. Indeed, the first conflict (in 1995) consisted of fierce opposition by populations in network expansion areas to the very high connection charge requested by the water company (in agreement with its concession contract). And the second conflict (in 1998) similarly involved already connected groups who refused to pay for the expansion of the network. It would seem that these conflicts reveal the antagonistic interests of connected and non-connected groups, as well as the deeper social significance of being connected, both in terms of user rights and in terms of social inclusion. However, as Schneier-Madanes shows, those conflicts were rooted in a context in which the population's initial support of privatization reforms (in many utility and public services) was progressively undermined by rate increases for many public services, the lack of subsidies to low-income families, and the new commercial character of the services. This situation was exacerbated by the impoverishment of large parts of the population in the context of a broader economic crisis. The socially inclusive properties associated with the connection to utility networks are hampered by the risk of being disconnected faced by a growing part of the city's population; and being *dis*connected may well be more stigmatizing than *not being* connected in the first place. This corroborates a conclusion of many studies: the physical connection to a centralized or "bundled" network is not by itself the ultimate solution to problems of access to essential services.

In a different institutional context, Marie Llorente discusses the reasons for the poor performance of the formally "bundled," publicly owned and operated water supply system in Delhi, and how this performance might be improved. Delhi's water supply is characterized by an insufficient and low-quality resource, poor condition of infrastructure (with massive wastage), insufficient and intermittent supply, strong disparities in access to and consumption of water, which affect low-income users disproportionately, and have poor cost recovery. Llorente includes in her discussion the many facets of this situation: rules, both formal (laws, policy, judiciary) and informal (customs, norms, codes of conduct); operators' internal governance structure (including incentives, the degree of bureaucracy, the level of autonomy and skills of agents and their behavior, and the representation of public interests); and users' practices and involvement (or the lack of it). Focusing in the conclusion on the question of whether or not a centralized network would be a sustainable global solution to Delhi's water problem, she argues in favor of a more diversified, demand-oriented approach, integrating centralized and decentralized supply via public or private providers. In particular, setting up a single, all-Delhi franchise contract would be a mistake because it would not provide a relevant answer to the two main issues: the fragmented nature of the city and the limited ability of customers in poor areas to pay. Thus, despite the unsustainability of current decentralized forms of water provision in Delhi, Llorente argues that a decentralized system should be preferred to a centralized, "bundled" network.

The next chapter takes a broader perspective, encompassing the "three worlds" of water use patterns. Bernard Barraqué argues that the model of water supply and waste water treatment systems services that was developed in industrialized countries during the twentieth century may not be sustainable. In the US, the extremely high levels of water consumption are jeopardizing the entire system. In Europe, the proliferation of environmental directives (laws) and of liberalization reforms in public services (based on so-called "full cost recovery") led simultaneously to more instances of non-compliance to drinking water standards and to larger water bills. The ultimate result is customers' growing distrust of their water utilities. In developing countries, Barraqué further argues, public–private partnerships and the privatization of services will not help to universalize access to water networks. In all contexts, sustainable services (economically and environmentally efficient services at socially and politically acceptable prices) will require "cheap money," public subsidies and cross subsidies, as well as the transition to "environmental engineering": resources protection, demand management, and, in specific contexts, alternative forms of service supply. But, Barraqué asks, "Who would want a 'substandard' septic tank in their garden when everybody tells them that networked-based and public sewage collection and treatment is the only real solution?" This question echoes discussions in other chapters on the (rhetorical and social) power of the notion of universalization and its sometimes questionable economic, social, or environmental benefits.

As a conclusion to this collection of chapters on challenges to, or associated with, network universalization, Gene Rochlin speculates about the deregulation or, as he puts it, the " de-institutionalization," of network-based large technical systems in the US and elsewhere. Viewing LTSs as social institutions in their own right, Rochlin calls for an analysis of reforms in LTS industries that goes beyond the dominant analyses of the preconditions, forms, and economic consequences of these reforms. He systematically explores the social impact of reforms on companies' staffs, the general population of "users" and public institutions. He argues that, as deregulation has:

> broken what were claimed to be their visible chains, humans are led to deny the costs, and the more insidious means by which they increasingly become technically, economically, and socio-politically bound by the means and mechanisms of "free market" rules, structures, and coordination requirements.

In many cases the break-up of large technical systems under the guise of deregulation did not lead as promised to the emergence of effective competition and competitive markets. Instead, those with the greatest or most effectively used market power are moving to re-aggregate the system, but this time largely free of the regulatory and government controls that restrained them from exploiting either their customers or their workers.

Concluding Remarks and Future Directions

Chapters in this book confirm the society-shaping (and not only socially shaped) nature of network systems and networking technologies. Networks are at the same time: socially and politically acknowledged standards of service (network-supplied basic services); socially normalizing devices; fundamental elements of the functioning of "network-dependent" societies (dependence can be assessed, in particular, through the cost of non-access or of network failures); a metaphor, even an "ideology"; and social institutions. As such, the chapters in this book are a contribution to bridging the gap identified by van der Vleuten (2001) between strong claims by scholars as to the major social importance of large technical systems and studies that focus on the internal workings of those large technical systems rather than on their interactions with society. The chapters do this in a way close to the "pluralist approach" to network studies that van der Vleuten advocates. By confronting very contrasted situations in different parts of the world, the book suggests future areas of research. In closing this introduction, we would like to emphasize three directions in network research that, in our view, deserve particular attention.

Networks, Cities and Spatial Dynamics

As noted above, the relations between technological networks and cities are crucial in two ways: first, because cities have, since the middle of the nineteenth century, been increasingly dependent upon the smooth functioning of a host of technological networks; and second, because cities are the loci and the foci of technological, economic, and social innovations that sustain the initial development of networks. The supply of networked-based urban services is usually a major responsibility of local governments and, therefore, an indicator of the ability of local governments to act. Focusing on the provision of network services gives essential insights into the evolving forms of urban government and into their capacity to implement public policies (see Le Galès and Lorrain 2003). Because network services are so central to the life and government of cities, the co-evolution of urban networks and urban spaces/societies discussed in several chapters of this book should be further explored, both empirically and theoretically. But this spatially sensitive study of networks should be fundamentally multiscalar, because networks intimately articulate spatial scales.

Networks and Sustainable Access to Essential Services

The "urban transition" that has affected (or is affecting) most areas in the world, resulting in an ever-growing portion of the Earth's population living in cities and in the multiplication of very large cities, was made possible by network technologies. Large agglomerations of population would simply not have been possible without the water, sewerage, transport, and energy supply systems carrying vital fluids to those cities and mephitic waste away from them, and connecting those agglomerations with their hinterland and with the rest of the world. But this does not necessarily imply that the contemporary forms of development of networks are sustainable economically, socially, and environmentally. The chapters in this book raise serious reservations. First, the cost of ever-stricter environmental and health regulations, together with full-cost-pricing and polluter-pays principles, may threaten the affordability of water supply even in the richest countries, not to mention the very serious problems faced by poor populations worldwide. Second, supply-oriented policies have favored the development of levels of consumption (of energy, water, cars, etc.) that are generally regarded as unsustainable, especially if they were to be extended on a world level. But the transition to more sustainable levels of consumption in advanced economies, and the control of rising levels of consumption in the rest of the world prove particularly tricky. Third, network-based supply has become the norm in the eyes of populations and of politicians, irrespective of its cost, its social accessibility, and its environmental efficiency. A crucial issue is thus: to what extent can reasoned use of networked and alternative forms of service provision be part of a sustainable scenario of universal access to basic services (and rights)?

Networks as Institutions

Social and political institutions tend to be taken for granted in studies focused on the development of technological networks in developed countries during stable policy phases. As suggested by the chapters in this book, however, when one broadens the analysis, geographically (to include developing countries) or historically (to include the less stable periods in terms of networks expansion, for example, the late nineteenth century and the recent period of liberalization/deregulation), the picture is quite different. In this book, the electricity crisis in California, conflicts over water supply in Latin America, or the pending crisis in the European model of urban water services, provide starting points for discussions of the interdependencies between the development of networks and the stability of social institutions. But the analysis can and should be taken further. A comparative approach seems very helpful to investigate networks as institutions and, even more fundamentally, as one in a group of alternative institutions structuring societies. What other "structuring institutions" did networks replace historically in societies where they are presumed to have taken on that function? What other structuring institutions exist in contexts where networks remain marginal (if only in terms of the proportion of the population connected to them)? These institutional issues may turn out to be key directions of investigation (in research as well as in policy terms) when searching for social responses to the vulnerability of advanced societies deriving from their dependence upon networks.

References

Braun, I. and Joerges, B. (1994) "Second order large technical systems," in J. Summerton (ed.) *Changing Large Technical Systems*, Boulder, CO: Westview Press.

Edgerton, D. (1998) "De l'innovation aux usages: dix thèses éclectiques sur l'histoire des techniques," *Annales Histoire, Sciences sociales*, 53(4/5): 815–37.

Graham, S. and Marvin, S. (1994) "Cherry picking and social dumping: British utilities in the 1990s," *Utilities Policy*, 4(2): 113–19.

—— (2001) *Splintering Urbanism. Networked Infrastructures, Technological Mobilities and the Urban Condition*, London: Routledge.

Le Galès, P. and Lorrain, D. (eds) (2003) "Gouverner les très grandes métropoles: Institutions et réseaux techniques," *Revue française d'administration publique*, Paris: La Documentation française.

Tarr, J. and Dupuy, G. (eds) (1988) *Technology and the Rise of the Networked City in Europe and America*, Philadelphia, PA: Temple University Press.

van der Vleuten, E. (2001) "Etude des conséquences sociétales des macro-systèmes techniques: une approche pluraliste," *Flux, Cahiers scientifiques internationaux Réseaux et Territoires*, 43 (January–March): 42–57.

Part I
Networks and the Development of Cities

CHAPTER ONE

Gig@city: The Rise of Technological Networks in Daily Life

Dominique Lorrain

This chapter analyzes the diffusion of network technologies as a complex and contingent process. In so doing, it participates in the debate on the "post-Fordist" city and the so-called tendency towards fragmentation and segregation (Castells 1997; Graham and Marvin 2001).[1] Our central hypothesis is that it is necessary to examine each element of a city's social structure before asserting such a trend (Harvey 2000; Marcuse and van Kempen 2000). In simple terms, the logic underpinning urban productive sectors is not necessarily the same as that governing housing markets, or that structuring the economy of technical networks. Society is constituted of many sub-parts which are, of course, interconnected and respond to the same general logic (globalization, market forces, etc.); however, I believe that it is erroneous to consider that a phenomenon on one level – if we use the metaphor of "instances" (or levels) coming from structural analysis – explains what is happening on another level (Lorrain 2001). It is necessary to develop analyses that seriously take account of the "built" (or technical) dimension of cities. This dimension is not a metaphor, it is a central part of how cities are organized, structured, and governed.

This chapter is divided into three parts. The first part argues that we are experiencing a new, third phase in urban history characterized by the growing role of technological networks and other infrastructures as key elements of modern life; the built environment structures the environment in which human beings live more than ever before. In the second part, the process of the diffusion of these networks is discussed, based on a number of case studies (water supply and the automobile). This process never begins as a universal phenomenon; in their infancy networks are limited to specific areas and players (individuals or enterprises). The third part analyzes the lessons of the process of generalization by considering two issues: the "splintering urbanism" argument and regulation activity. If we consider the central role of the built environment and recognize that it is the product of human action (individuals, organizations, institutions), then the organization of the built environment must be seen as representing new responsibilities for governments. This certainly involves greater responsibility than the classical regulation of national monopolies. The division between utilities and the services provided by them are blurring; in some industries the services provided via "the pipe" are more strategic than the pipe itself. This raises the question of exactly what should be regulated.

Three Types of Cities

Taking the example of Tarr and Konvitz (1981), we propose to discuss three historical types of urban structures: the city, the megalopolis, and the gigacity (Lorrain 2000a: 12–13). The transition from one type to the next is not only characterized by growth, but also by a change of pattern.

Polis: *The City of Pedestrians*

The first type of city, the *polis*, has a long history spanning many centuries. It is characterized by some general features: spatial focus around several public buildings (palace/fort, church/temple), low buildings, public fountains and individual wells, and limited transport techniques (Mumford 1964). The *polis* was a distinct territory from the rural area under its control. It had commercial dealings with its surrounding rural environment and with other cities or states, which were sometimes very far away. Nevertheless, the city was mostly a closed space surrounded by walls and organized around a citadel, a fort. In his seminal history of cities, Lewis Mumford accurately describes the different phases in this long period of the pedestrian city: antique, middle age and baroque (Table 1.1). In these cities, spatial relationships were determined under a principle of contiguity (relations with those who are close). The value of the fixed structures (networks) that framed the city remained low in comparison to the value of other elements of the built environment (palace, fort, cathedral, etc.).

The Megalopolis

The break with the *polis* began in the nineteenth century with the development of new sources of energy, railroads, and the first capital intensive underground networks: water, sewerage, and subways (Tarr and Konvitz 1981). After the

Table 1.1 *Basic Elements of the Pedestrian City*

Period	*Type*	*Outstanding Buildings*
Ancient city	Citadel	Palace, grain warehouse, temple
Medieval city	Walled	Fort, cathedral, cloister, hospital, market place
Baroque city	Palace	Treasury, prison, avenue

Source: Adapted from Mumford (1964), chapters 9, 10 and 12.

Second World War, changes in building techniques (the widespread use of concrete and the massive use of glass frames), the development of elevators, and the mass diffusion of automobiles led to a new form of urban center: the megalopolis. Cities expanded both horizontally and vertically. This was the era of skyscrapers, the most advanced types of which were found on the East Coast of the US. The density of networks changed, and cities were reshaped by these "first" heavy networks and by automobiles.

The Gigacity

At the beginning of the twenty-first century, we are experiencing a new type of city characterized by several features. The first salient feature is size (see Table 1.2).[2] In 1960 there were two megalopolises with more than 10 million inhabitants; by 2000, there were 20. The second feature of the changed city is the expanded role of networks. A gigacity can be defined as a place with a high density of networks. Gabriel Dupuy mentions that in the modern city, reshaped for use by automobiles, the dense network of streets represents "up to 30 percent of the urban surface, in Los Angeles 40 percent" (Dupuy 1995). Water and waste water mains, electricity networks, and mass transport systems are ubiquitous; they are the first level of modern cities. New networks, including telecommunications and cable networks have been added to cities, as have new techniques

Table 1.2 *The Twenty Largest Cities in 1960, 1996 and 2015 (projected) (in millions of inhabitants)*

City	*1960**	*1996*	*2015*
Tokyo	10.4	27	29
Mumbai	4.5	16	26
Lagos	0.7	11	25
São Paulo	3.2	17	20
Dhaka	0.6	9	19
Karachi	1.9	10	19
Mexico City	3.1	17	19
Shanghai	6.9	14	18
New York	11.3	16	18
Calcutta	4.6	12	17
Delhi	2.6	10	17
Beijing	4.0	11	16
Manila	1.1	10	15
Cairo	–	10	15
Los Angeles	6.5	13	14
Jakarta	2.9	9	14
Buenos Aires	7.0	12	14
Tianjin	3.2	10	14
Seoul	3.0	12	13
Istanbul	1.5	8	12

Sources: Beaujeu-Garnier *et al.* (1966) for the year 1960, UNDP for 1996 and 2015.
Note: *Depending on the country, the year of the census can vary from 1958 to 1963.

for movement, such as elevators. Third, a new dimension was added to cities with the development of underground spaces and high-rise buildings.[3] A fourth element is the way in which new networks are changing the relationship of cities with the rest of the world. With fast trains, the generalization of air travel, and the diffusion of cable networks and the internet, the city no longer has any boundaries. This change marks a shift from the old principle of contiguity to a new principle of connectivity (Offner and Pumain 1996). A fifth aspect concerning new developments in the built environment and technical networks is the parallel diffusion of robots, i.e., technical devices used in daily life which are complementary to the network (e.g., the addition of GPS systems in cars). These new developments in urbanization move cities from the realm of megalopolis, corridors, or urban regions, characterized by sprawling, urbanized spaces, to that of an archipelago, where cities represent islands concentrating activities and exchanges (Veltz 1996). The techniques of exchange then become strategic within an economy of flux and nodes.

The Process of Diffusion

Inhabitants of modern megalopolises and gigacities take the built environment for granted with its complex mix of buildings, public equipment, technical networks, and mechanical devices. However, this environment is the result of a long process of development and diffusion and it requires intensive control and maintenance. The development of networks and their diffusion into daily life have not followed a rational pattern – they have evolved through trial and error.

The Infancy of a Network

In the first phase (infancy) of a network, few people are connected and the service is expensive. During these early years, the well-off have access to these symbols of modern life: running water, electricity, telephones, or private cars. In general, it is a time of experiment and invention; several technologies are competing (as was the case with electricity): the system has not been stabilized.

Water supply, at the turn of the nineteenth century in France, was limited to public fountains. Water mains had been laid in the center of major cities, but the periphery (suburbs and villages) had no access. In the city of Lyon, described by Franck Scherrer, the first contract to improve water distribution was signed in 1853 with the newly established *Compagnie Générale des Eaux*. Ten years later, 10,000 households had been connected; this number rose to 20,000 after 20 years, which represented only one-sixth of all residents. Water was not easily available. Under the initial agreement, the daily production was 20,000 cubic meters, half of which were for municipal requirements (street cleaning, public fountains, etc.). Five years later, these needs were estimated at a minimum of 45,000 cubic meters (Scherrer 1997: 49). Other cities had similar experiences. In 1841, Bordeaux had 120,000 inhabitants and provided the equivalent of 3.5 liters per person per day of running water. As noted by Jean-Roland

Barthélémy, "the municipal effort began in 1854, however, it was only in 1880 that a municipal water agency was created" (Barthélémy 1997: 63).

One century later, the same situation can be observed with regard to another modern network: the cellular phone. In the mid-1980s, this new technology was only in the experimental phase in France. Only one network was available, under the responsibility of the incumbent state enterprise. This was poorly diffused, expensive, and not very convenient (the calls were transferred through an operator). When a second license was granted in 1987 (after a process of direct negotiations between the French Ministry of Post and Telecommunications and the *Compagnie Générale des Eaux*), the cost of equipment exceeded US$3,000 and there were less than 30,000 subscribers to the state company's network. In 2001, just 14 years later, Cegetel, the CGE subsidiary was operating the second network. It has 10.1 million subscribers, and, for the first time in France, the number of cellular telephones exceeds that of fixed lines.

Expansion of the Network

Public policies have been central to expanding networks and increasing the supply of services. These policies establish priorities, determine norms, and design institutional frameworks that are of key importance in facilitating transactions in a sector where markets have been partially inadequate (North 1990). Many examples exist from developed and emerging countries. They demonstrate the central role of the state during the expansion phase of network development.

Even though the first French water corporations were created in 1853 and 1880, compared to other industrial countries, France lagged a long way behind other developed countries. This was due to two factors. First, a cultural attitude of indifference to hygiene that was mentioned by many observers, reformists, and members of visiting missions. Second, an inadequate institutional framework. At this time, water was considered a "public good" with low tariffs; the levels of capital expenditure were low, with most resources coming from state grants. Consequently, the development of water mains was dependent on state budgetary policies, which had their own logic, resulting in intermittent development. This was a major weakness in the development of a water policy. It took until the 1930s to develop a new framework. The basic change that occurred was the realization that water had a cost, so meters were installed, tariffs were raised, and utility companies or municipalities began to generate positive cash flows that were then allocated in order to modernize the network.[4] After the Second World War, in 1954, an additional tariff was assigned to a specific fund dealing with the modernization of rural areas in order to make the service universal (the same mechanism had previously been set up for electricity). The process of expanding the water networks continued and by the mid-1960s, French backwardness in this area was a thing of the past.

In another country, Argentina, specifically in Buenos Aires, we have a case where an inadequate tariff structure and poor management hampered the existing water company, *Obras Sanitarias de la Nación* (OSN), which was among the best water companies in the world before the Second World War (Dupuy

1997). When reform was instituted at the beginning of the 1990s, the position of this company had seriously deteriorated since its halcyon days (Faudry 1999; Schneier-Madanes 2001: 45–63):

- the organization was over-staffed and this had adverse consequences on costs;
- due to a poor commercial policy, only 75 percent of bills were paid;
- the need for treated water was considerable due to leaks and illegal connections. There was a discrepancy between the amount of water produced and the amount sold – this was estimated at 43 percent ("estimated," because there were no meters on many parts of the network, and accurate figures were difficult to obtain);
- there was a generous flat rate for the middle class and the rich living in the city center. These sectors consumed an average of 573 liters per capita per day (lcd) for all the networked area, but 700 lcd in the *Capital Federal* (the city center).

In 1993, this situation led to a privatization reform. Nine million people were included in the service area of the contracting firm; however, at that time only six million were connected to the water network. This meant that the extension of the network was essential to the success of privatization efforts. But expansion was made difficult by an inadequate institutional framework (tariff setting). The concession agreement provided for connection charges (*cargo de connection*) and a charge to finance the new network (*cargo de infraestructura*) for both the water and the wastewater networks. The cost was 1,455 pesos per household, compared with an average monthly income of 240 pesos. The users had to pay the real cost of connection (principle of marginal cost pricing); this represented a charge of six months' income. After the first extension program in 1995, the users refused to pay these charges. They demonstrated; opponents to privatization resumed their criticism. The issue rapidly became politically sensitive. The entire process was stopped and new mechanisms had to be established. The principle used was the consideration of the average cost as a way to calculate the charges. All users (those already connected and those soon to be connected) had to pay an additional charge called the *SUMA* (*servicio universal, mejora ambiental*) which could not be more than three pesos per month. Those soon to be connected had to pay an extra charge of 2 pesos per month for each service, for a period of five years; this represented a total of 120 pesos (Faudry 1999, Schneier-Madanes 2001, and this article). Again, the role of the public sector in the design and legitimacy of new rules was central to the development of this network.

China offers another instance of a clearly defined, strong institutional framework based on the public sector that has been relatively efficient if we consider the pressure of the needs (Sogreah Consultants 2000). In the period from 1990 to 1998, water treatment capacity and the size of the distribution network in China have increased by almost 50 percent, from 140 million to 210 million cubic meters per day, corresponding to a total investment of ¥79 billion ($9.5 billion). Much of the investment in the water sector came from government funds (either national, provincial, or local). In mid-2000, "only 24 operations with the private sector have been identified; they represent approximately $580 million or 6 percent of the investment in the water supply sub-sector" (Sogreah Consultants 2000). There is nothing extraordinary in this achievement and in the ability of the Chinese authorities to oversee the development of their infrastructure. They have set up a framework that provides the basic answer to their problem: a well-structured municipal government has proven to be an efficient means of managing the extension of water networks in urban areas. They have installed meters (90 percent of the urban population has access to tap water). As in many developing countries, the original policy was to have low tariffs. This policy was reconsidered in the 1990s and water companies increased their rates and, consequently, their resources (Lorrain 1998: 5–21).

India, with similar quantitative issues, provides a very different example. Part of the problem is rooted in a poorly designed public policy with many elements similar to the Buenos Aires example. Insufficient control of the urban process and low tariffs have resulted in a lack of funds and a deterioration of the service, both in terms of the quantity of water available and the quality of the tap water. "Cities are facing many problems and demographic growth is only one of these. There are numerous other constraints: an inefficient infrastructure, badly designed urban regulations, weak municipal institutions, and inadequate financial services and funding for urban development" (Zérah 2000: 16). The central point to consider is the reaction of households and the related consequences for the future development of the network. Households have reacted to the poor level of municipal service; they make their own decisions on how to improve their situation. They dig wells on their own property; they invest in motor pumps, storage tanks, and, frequently, in simple treatment devices. Overall, this private investment in equipment has a cost. "The unreliability of water supply costs Delhi Rs 3 billion annually; this is twice the municipal expenditure on water" (Zérah 2000: 144). Furthermore, this investment in private equipment impedes the future development of the network. If the public system was upgraded and if every household had access to tap water, this investment in equipment would become superfluous and would represent a loss for the household. This issue of modernization is sensitive because those who have invested in such equipment are not the poor; they have the political ability to express their grievances to elected officials. The Indian case thus provides an example of a "vicious circle" regarding the role of institutions. The originally bad institutional framework prompted households to seek private solutions which, in turn, hampered the development of water mains and distribution pipes throughout the city.

Generalization: Networks and Services

The third step in the diffusion of networks is characterized by increasingly frequent practices. At this stage of "extension," a clear distinction has to be made between the network and the "practices behind the meter." The extension of the user's practices and the diversification of policies of network companies (from pipe to services) leads to the disappearance of the barriers between "before" and "after" the meter, or the difference between the network (the regulated utility) and private practices. The point we want to stress here is that the present-day gigacity is not only characterized by a growing number of networks and of uses of networks, but at the same time by the development of various goods that are extensions of these networks. This socio-technical mix of utilities and private equipment complicates regulation and changes the role of public action. We will illustrate this point with the example of the diffusion of the automobile (a similar demonstration could be provided for the diffusion of electricity, natural gas, or the telephone in modern daily life).

The diffusion of the automobile has led to the development of roads, highways (two networks), and other complementary facilities and services: gas stations, repair shops, rest areas, car parks, and insurance. The diffusion also depends on a car industry with sophisticated strategies for selling its products (cars) to the greatest number of consumers. In such a case, the process of diffusion combines three elements: the network, the product (the car), and the education of drivers.

The network. The investments to develop roads, increase their length, and improve their safety have been a permanent concern of European public policy since the Second World War. "It is now a fully integrated system . . . which goes beyond institutional boundaries" (Dupuy 1995). Some rules, which aim to organize individual practices on the network, have been diffused all over the world. The first red light goes back to 1914, as does the first stop sign posted in Chicago. As Gabriel Dupuy writes, common standards mean that a motorist "can drive his car across borders knowing that there will be no surprises on the roads in other countries" (ibid.).

The product. The same standardization process has occurred in the car industry. Everything has been done to simplify the driver's job,[5] to provide a similar local environment (in the car, the gear box, the buttons for the lights, the horn, etc.). The tendency is towards simplification and convergence; the result is the homogeneity of the fleet.

Education of the driver. At the same time, policies have been established to educate future drivers. In the US, driving licenses were first issued in Chicago in 1898, followed by the State of New York in 1901; licensing was universalized throughout the country only in 1926 (Flink 1975; Boullier 2001). Rapidly, the car industry developed sophisticated strategies, not only in order to sell its products, but also to create the necessary general conditions for the business.

The product had to be desirable: marketing campaigns have associated cars and freedom. The car is a convenient tool for traveling, however, it depends on having a map of the network mentioning all available facilities: where to stay and what to visit (Baudant 1980; Karpik 2000: 369–89). This was the genius of Michelin; being the first company in France to understand that these elements were critical for the development of its tire manufacturing activity. It entered the guide and road maps business. This still exists today and the *Michelin Guide* is a must as regards hotel and restaurant classification.

Lessons

Time is a Central Factor

The process of expansion takes place over a long period of time (several decades); the first phase of construction is followed by expansion of the network and then by the diffusion of practices. This general process, consisting of three phases, leads us to question the well-established thesis of segregation in technical networks, especially if the operators are private firms. Briefly, the segregation argument can be developed along two lines: (1) firms only have an interest in the wealthy (cherry picking attitude), and (2) this results in a divided city (splintering thesis) (Graham and Marvin 1994: 113–19; Guy *et al.* 1999; Graham and Marvin 2001).[6]

Most of this thesis is not supported by facts; a large part of it is based on the housing market and on telecommunications. Starting with the telecommunications network, the authors generalize to other networks, however, they do not consider whether these technical networks had different technical characteristics and different histories and, whether, because the development of telecommunications is relatively recent, it may not necessarily be relevant to all technical networks (Lorrain 1995: 47–59). In the examples we have previously mentioned, one common trend can be observed. The expansion of these networks began with a small group of users; this was the case for automobiles and the early use of electricity, gas, and water. In the beginning, these technologies were expensive; sometimes they were not easy to adopt and required major modifications and personal investment of time (e.g., the case of automobiles before 1960 in France, or the first years of microcomputers and the internet). The development of "heavy" networks always starts in and develops from the city centers; they are dedicated to those who can afford them – businesses and the rich. Three factors then converge to produce a generalization of the networks and mass diffusion.

Public Rules

The large urban networks we have been discussing have not traditionally been organized under free market rules. They have been considered "public utilities," or "public services"; new categories of "universal services" and "services of general economic interest" seem to be emerging in Europe, but still have not been clearly defined. These networks have been considered essential to society and, therefore, too important to be left to operate under the laws of supply and demand. Their organization has been guided by several principles, one of which is access for all. We have no clear evidence to demonstrate that these principles have been abandoned. On the contrary, we have the example of water and electricity supply in emerging countries. Changes were introduced by pro-market reformers; however, the contracts signed with the private sector have always included an obligation to expand the networks. The achievement of this goal can take time; it can sometimes be chaotic, but it means that these pro-market reformers hope for universal access and not segregation. Nevertheless, if segregation does occur and if "social dumping" becomes a reality for some groups and for some neighborhoods, this is less a specific feature, inherent to the technical networks themselves, than a product of society in general. This would mean that the public in a particular country has accepted tariffs and rules for connection that will produce differences among inhabitants. If this occurs, the society would also be fragmented in other sectors – housing, education, transportation, labor, etc. In such a case, technical networks merely reflect the previously accepted view of a divided society.

The Logic of the Firm

It is an over-simplification to consider that the goal of a private company is only "to maximize profits for shareholders"; this is the debate between "maximizing" and "satisfying" (Simon 1979; Winter 1991). Private utilities have to serve five stakeholders: consumers, the regulatory authorities, their shareholders, employees and managers, and the firm itself. From one country to another, from one sector to another, from one time period to another, the equilibrium between these five forces may change (Chandler 1994; Crouch and Streeck 1996). This is apparent in the recent history of the electricity industry. At the end of the 1980s, when the movement towards deregulation began, a firm's profits could be increased in three ways: by a technology push, a reduction in the labor force, or through a shift from coal to gas (or nuclear power) as a source of primary energy. In the case of the UK, a direct confrontation with the miners, a constant reduction in employees, and a shift towards small gas turbines rapidly generated profits. However, because of the dynamics (some would say aggressiveness) of the financial markets, a large portion of these profits were distributed to managers and shareholders (Glachant 2000). In Germany, the combination of these factors was different both in the way profits were generated and in how they were allocated. First, the large electricity companies agreed to pay an extra charge to manage the transition to coal; they had been present for over a century in the coal producing regions and recognized that they had responsibilities. A large part of their efficiency was achieved through better vertical integration. Second, the allocation of profits between the various stakeholders was more balanced.

We could make the same observations with regard to the large French utility groups, Suez and Vivendi. These two companies never considered profits to be the central aim of their policy.[7] They had good reason to think that way. In a market where they have to compete for access to contracts, a central factor in their development is satisfying their clients – the local municipality (which is the organizing authority and signs the contract) and users. Of course, these companies have a monopoly, but not an absolute one. In particular, this monopoly is reviewed periodically. Therefore, it is risky for them to be involved in conflicts, either with their employees or their customers. UK water companies provide a different example – they used their profits to offer their top managers very high wages. As a result, their image and reputations were badly damaged during the 1995 drought. This led to political criticism, to a questioning of the model of reform, and to a considerable increase in their obligations under the "final determinations" published in 1999 by the Office of Water Services (OFWAT), the water industry regulator.

What we have learned from the most successful private companies that have been established in these sectors for a long period, and which are still expanding, is that a company's reputation is a fundamental asset. Indeed, happy users make for happy companies. In order to achieve this, managers of these companies know that they must establish reasonable policies, demonstrate technical proficiency, and deliver affordable services. They also understand that it is not possible to operate a service in a city where a large and permanent part of the population is excluded. This could create an explosive situation, which is a major risk for a company as the denial of basic utilities by a private company can have dire political and economic consequences – civil disturbances and loss of a contract.[8] In other words, while the logic of these firms is obviously not to lose money; they are not eleemosynary institutions. Their goals are to expand over time, to obtain new contracts, and to satisfy clients. Ultimately, they wish to reach every household; there are no good or bad consumers, there are only people who consume services and pay their bills.

Telecommunications companies are somewhat different because they can be seen as both utilities and commodities. As providers of commodities, different companies compete for "niche" markets, especially those of businesses and wealthy consumers. However, this dual nature of services cannot be applied to other utility networks such as water and wastewater, solid waste, and electricity (even though, in this last case, consumers in some countries may choose their distributor; in such a case, the splintering of supply does not necessarily mean discrimination and segregation of services).

Club Effects and Economies of Scale

A third property has to be taken into account in a discussion of segregation in the network industry. For most goods and services, the satisfaction of a consumer is independent of the overall number of consumers (Chandler 1994). For network services, however, this is different. If a telecommunications network has only one cable and two clients (the same applies to a road network) the number of possible connections is low, as is the social utility for the two pioneers. Therefore, it is easy to understand the notion of the "club effect": the greater the number of people connected, the greater the number of possible connections and the more valuable the service. Additional users increase the advantages of being connected. This property, specific to networks, creates a "demand pull" which expands the network and increases the number of people connected. This process can be accelerated by a technology push, as was the case with the cellular telephone industry in the 1990s or the micro-computer sector. It is reinforced by economies of scale: the greater the number of consumers, the smaller the share of sunk and other fixed costs (commercial costs involved in obtaining the contract, research costs to improve the technical process, etc.) borne by each customer. These two phenomena (club effects and economies of scale) combine to expand the network. If we understand the lessons of history, successful networks (networks that have not disappeared) are those that have expanded beyond their initial niche (either socially or spatially defined) to encompass the whole population, without excluding any group/section.

For all these reasons the segregation thesis is inadequate for the purpose of describing the history of urban networks. If discrimination can be observed, this means that we are in the infancy of the network and that it is organized under a dual regime (utility and commodity), or that the society as a whole is fragmented; in such a case, the situation is not rooted in the public policies that organize the technical networks, but in more general factors. In this case, researchers have to investigate political failures in society and in the establishment of democracy. The ultimate goal of networks and firms is expansion and diffusion. The idea of *fluidity* usually associated with networks is therefore a genuine concern in the provision of network services. Networks carry flows; the ultimate goal of an operator is to achieve a totally fluid flow: no disruption, no accidents, no conflicts. It is to deliver a good or a service to everybody, 24 hours a day, 365 days a year, trouble-free. Following such logic, a permanently segregated society would be too risky and would not necessarily lead to greater profits.

Expansion and Diffusion

The diffusion of networks is based on the expansion of the networks as well as on the "social practices behind the meter"; this implies two sets of players. In the beginning, technological networks (referring to the notion of utility) are clearly separated from private goods and from the private sphere (domestic activities). These technical networks are subject to specific regulations; they are organized under a monopoly and operated by large public or private corporations. The first phase of expansion is characterized by the stabilization of the networks themselves. We have discussed the socio-political construction of networks as a stabilization process consisting of four elements: technologies, institutional architecture, rules and norms, and values (Lorrain 2000b). The process is largely influenced by public policies and the role of administrative bodies; in their infancy the development of networks is concerned with the so-called "system builders" (Hughes 1983). During this strategic period in the infancy of a technology, these inventors, engineers, and administrators play a primary role in stabilizing all the elements that enable the early diffusion of the system.

If we ask the question, "Why do some networks grow and spread among users?" we see that success is not linked only to public policies. A wide spectrum of social practices also has to be taken into consideration. Networks expand because they satisfy global needs. They act as symbols of progress, offering access to a healthier lifestyle (water, electricity) and personal freedom (the car, the telephone). Their diffusion is accelerated by the fact that these networks provide the basic support for a wide range of practices. These practices are characterized by the utilization of goods and equipment delivered and promoted by private firms. In such a case, diffusion is not only the result of the strategy of the network companies (utilities want to promote electricity), but it also results from a more complex mix of practices based on these facilities and the supply of many goods that satisfy the needs promoted by the industry as a whole. The changes in the kitchen over a century and the expansion of the use of small appliances, based on the consumption of electricity, are a good example (Cowan 1983). As a consequence, electricity has become an essential commodity because it is the energy source on which all these kitchen appliances depend. This is the same "system of mixed private–public economics" that has characterized the automobile and contributed to its success over the last century.

The effects of such a diffusion raise the issue of the limits of the networks and also of what needs to be regulated. The boundaries between what was considered public (networks) and private (domestic) have changed. If we accept that utilities expand after the meter, then there would be no limit to the notion. In such a situation, one could state: a utility encompasses not only the facility provided (e.g., electricity) but also the private practices based on that facility (e.g., cooking). If so, every practice would have to be regulated, which is impossible. In the case of water and electricity, the support facilities and the service are jointly operated by the same company. With telecommunications and the e-economy, support facilities are considered a utility and the service a commodity. However, this raises the following question: what is the most strategic, i.e., the most valuable part: the pipe or the flux?

States must address how, or even whether to regulate this service. In the past, one kind of regulation provided for universal access. In the case of telecommunications, for example, this might mean that carriers must provide a certain level of connection at a minimum cost. But what is the equivalent level of service in terms of access to the internet? What kinds of information should be considered essential and, therefore, to be accessed by everyone as a matter of course (e.g., information on the best providers of other utilities – water, gas, electricity)? Also, is it necessary to provide access to this service in the home (as is the case with water and electricity) or can this service be provided in public locations?

As with the regulation of access, there are many content-related issues. With water and electricity, this was a relatively simple matter. Standards were set to define the potability of the water delivered to the end-user and the voltage level of the electricity supplied. With the internet and cable networks, the regulation of what is delivered is much more problematic. Should the state regulate the delivery of pornography, racist messages, anti-democratic screeds, the promotion of Nazism, etc.? (The US constitution makes many of these questions a moot point. Apart from child pornography, there is little content that can be regulated.)

Another issue raised by the new capabilities of utility networks is that of the level of the protection of privacy. Because of the infusion of information technologies into all networks, the potential now exists to accumulate a large amount of data on each individual using a network. Of course, in the past, utilities had access to a considerable amount of data on a household's consumption practices, but only through graphs of the peaks and volumes of consumption habits. This data, however, was seen as basic information necessary for the efficient operation of the system (i.e., calibrating supply and demand). In the old utility culture, this information was not viewed as a commodity that could be sold – or that could be used to sell something to customers. This is now the case,[9] and this raises questions about privacy and the need to regulate how this information is used.

Let us illustrate this point. Several years ago a court case in France showed how much information could be gathered on someone to establish his whereabouts at certain times. During a case concerning bribery, a judge tried to ascertain the truth of an accusation that the manager of a soccer team, who was also the new French Minister of Urban Affairs, attempted to bribe the manager of an opposing team. The opposing manager asserted that he and the Minister had met in Paris on a certain day, at a certain time and arranged to fix the outcome of a game. A third party, the mayor of a city in Northern France, asserted that the meeting could never have occurred because the Minister had come to the mayor's city, and they were meeting in the City Hall of this Northern city at almost the same time, and on the same day as the supposed meeting in Paris.

The judge in the case ordered an investigation into the stories of all three men. The investigator discovered that the mayor had had a meeting in Paris on the morning of the day in question. Then, by checking his credit card records, his cell phone records, and the toll-road database, it was established that the only way the mayor could have been back for a meeting in his own city, as he had testified, was by driving from Paris at an average speed of over 100 mph. After months of investigation, the mayor admitted he had lied.

For our purposes, what is important about this anecdote is that it opens a debate concerning some of the technical tools of control available in modern society. It is not exactly "Big Brother," but it's getting close to it. In any case, this "invisible hand of control" is the other (more negative) side of socio-technical networks that have been built by the welfare state to protect and serve citizens. Thus, the diffusion of networks in modern society has the following two dimensions: easy access to many services and facilities *and* an invisible net generating data on what we buy, where we go, and whom we call. In the final stage of diffusion, when reductions in cost have made these services available to everyone, the question of privacy becomes central and should be the primary focus in regulating networks.

When a network has diffused to the point that its service is considered a necessity, regulations must ensure the reliability of that system. This is the point that symbolizes the success of a network – when it has become so widespread that it is seen as a basic element in modern life. As the number and the size of technical networks continue to grow, so does the complexity of coordinating all these systems. Our society knows how to organize and regulate individual networks, but its new challenge is to coordinate multiple systems that have evolved into mega-networks. In emerging countries, these new developments raise questions of institutional design, particularly in those metropolises where the deregulation of existing, vertically integrated utilities has led to the creation of very complex institutions.

In addition to raising questions about newly configured institutions, advances in network technologies also raise questions about regulation, especially in light of the shift in focus from infrastructure to service (from pipes to flux; from hardware to software). In general, the water system continues to be organized around the physical infrastructure. However, automobile systems and information and telecommunications systems are much more difficult to organize and to regulate. In the gigacity, when technology has reached an advanced stage of diffusion, the central issue is not "splintering" or the "digital divide." Rather, it is the level of control over individuals made possible by these technologies and the obstacles to the regulation of these networks.

Notes

1 See also the two following chapters in this volume for a continuing discussion of the "splintering urbanism" thesis.
2 Sixteen percent of the world's population was urban in 1900. Forty-five percent of the world's population was urban in 1990. In 2000, there were 320 cities of more than 1 million people, and in 2015 there will be 22 cities of more than 10 million people.
3 See the works of the French historian André Guillerme on underground urbanism.
4 For a more detailed history of these institutional changes see my chapter (Lorrain 2000b), pp. 167–71.
5 Gabriel Dupuy explains that the introduction of the electrical choke facilitated the diffusion of automobiles to women.
6 Even if the words are different, this is much the same argument as that shared by many radical urban sociologists: see Castells, Harvey, Marcuse.
7 Vivendi Universal (the better-known part of the former Vivendi group) differs in many respects from the environment services branch from which it originated, the *Compagnie Générale des Eaux*, now Veolia Environment. This is especially true in terms of profits and stakeholders' remuneration. Compared to the very high financial profits of Vivendi Universal, the study over a period of 20 years for both *Compagnie Générale des Eaux* and *Lyonnaise des Eaux* (the other French major groups) shows that the ratio between operating profit and turnover in these companies is reasonable, with a central value of 7.2 percent (see Lorrain 2003).

8 Something very much like this happened in Bolivia where a utility's bad pricing policy led to street disturbances, injuries, and deaths. As a result, the company, Cochabamba, lost its contract. Word of this incident circulated widely among utility companies worldwide. See de Gouvello (2001).
9 United Utilities, a British company, sold its electricity customer information data base in 2000 to TXU, an American firm. In the Philippines, Meralco, a supplier of electricity in Manila is now offering credit cards and other financial services to its 3.7 million customers (Landingin 2001).

References

Barthélémy, J.R. (1997) "Bordeaux, 1903–1993," in D. Lorrain (ed.) *Urban Water Management: French Experience Around the World*, Paris: Victoires-Hydrocom Editions.
Baudant, A. (1980) *Pont-à-Mousson (1918–1939): Stratégies Industrielles d'une Dynastie Lorraine*, Paris: Publications de la Sorbonne.
Beaujeu-Garnier, J., Gamblin, J., and Delobez, A. (1966) *Images Economiques du Monde*, Paris: Sedes.
Boullier, D. (2001) "Les conventions pour une appropriation durable des TIC: utiliser un ordinateur et conduire une voiture," *Sociologie du Travail*, 43: 369–87.
Castells, M. (1997) *La Société en réseaux*, Paris: Fayard (3 volumes).
Chandler, A.D. (1994) *Scale and Scope: The Dynamics of Industrial Capitalism*, Cambridge, MA: Harvard University Press.
Cowan, R.S. (1983) *More Work for Mother: The Ironies of Household Technology from the Open Hearth to the Microwave*, New York: Basic Books.
Crouch, B. and Streeck, W. (eds) (1996) *Les Capitalismes en Europe*, Paris: La Découverte.
Dupuy, G. (ed.) (1987) *La Crise des réseaux d'infrastructure de Buenos-Aires*, Paris: LATTS, Université de Paris XII, ENPC.
—— (1995) "The automobile: a territorial adapter," *Flux*, 21 (July–September): 21–36.
Faudry, D. (1999) "La concession des services d'eau potable et de l'assainissement de Buenos Aires," in D. Lorrain (ed.) *Retour d'experiences: six cas de gestion déléguée à l'étranger*, unpublished report, Paris: Ministère de l'Equipment, Isted.
Flink, J.J. (1975) *The Car Culture*, Cambridge, MA: MIT Press.
Glachant, J.M. (2000) *Les réformes de l'industrie électrique en Europe*, Paris: Ed. du Commissariat Général du Plan.
Gouvello, B. de (2001) "Les services d'eau potable et d'assainissement de Cochabamba: de la concession à la guerre de l'eau," in D. Lorrain, (ed.) *Experiences de Services Urbains 2*, unpublished report, Paris: Ministère de la Recherche.
Graham, S. and Marvin, S. (1994) "Cherry picking and social dumping: British utilities in the 1990s," *Utilities Policies*, 4: 113–19.
—— and —— (2001) *Splintering Urbanism*, London: Routledge.
Guy, S., Graham, S., and Marvin, S. (1999) "Splintering networks," in O. Coutard (ed.) *The Governance of Large Technical Systems*, London and New York: Routledge.
Harvey, D. (2000) *Spaces of Hope*, Edinburgh: Edinburgh University Press.
Hughes, T. (1983) *Networks of Power: Electrification in Western Society 1880–1930*, Baltimore, MD: Johns Hopkins University.
Karpik, L. (2000) "Le Guide Rouge Michelin," *Sociologie du Travail*, 42: 369–89.
Landingin, R. (2001) "Meralco prepares to exploit a new era in Philippine power," *Financial Times*, 21 June.
Lorrain, D. (1995) "The regulation of urban technical networks," *Flux*, 21: 47–59.
—— (1998) "Les réseaux techniques urbains à Chengdu: de l'administration directe à l'économie de marché," *Flux*, 33, July: 5–21.
—— (2000a) "Apres la mégalopole, la gig@city," *La Recherche*, December: 12–13.

—— (2000b) "The construction of urban services models," in A. Bagnasco and P. Le Galès (eds) *Cities in Contemporary Europe*, Cambridge: Cambridge University Press.
—— (2001) "Un livre extrême," in B. Lepetit and C. Topalov (eds) *La Ville des Sciences Sociales*, Paris: Belin.
—— (2003) "L'économie paradoxale des réseaux techniques urbains," in Henry Cl. and Quinet E. (eds) (2003) *Concurrence et Service Public*, Paris: L'Harmattan, pp. 163–94.
Marcuse, P. and van Kempen, R. (eds) (2000) *Globalizing Cities: A New Spatial Order?*, Oxford: Blackwell Publishers.
Mumford, L. (1964) *La Cité à travers l'Histoire*, Paris: Seuil.
North, D.C. (1990) *Institutions, Institutional Change, and Economic Performance*, Cambridge: Cambridge University Press.
Offner, Jean-Marc and Pumain, Denise (1996) *Réseaux et territoires, significations croisées*, La Tour d'Aigues (Fr.): éditions de l'Aube.
Scherrer, F. (1997) "Lyon: a century of history in public and private management," in D. Lorrain (ed.) *Urban Water Management: French Experience Around the World*, Paris: Victoires-Hydrocom Editions.
Schneier-Madanes, G. (2001) "From well to network: water supply and sewerage in Buenos Aires (1993–2000)," *Journal of Urban Technology*, 8(3) December: 45–63.
Simon, H. (1979) "Rational decision-making in business organizations," *American Economic Review*, 69, December.
Sogreah Consultants (2000) *China PPI Framework Initiative: Water Sector*, final report, Washington, DC: World Bank.
Tarr, J. and Konvitz, J. (1981) "Patterns in the development of the urban infrastructure," in H. Iette and Z. Miller (eds) *American Urbanism*, New York: Greenwood Press.
Veltz, P. (1996) *Mondialisation, Villes et Territoire, L'économie d'archipel*, Paris: PUF.
Winter, S.G. (1991) "On Coase, competence, and the corporation," in O.E. Williamson and S.G. Winter (eds) *The Nature of the Firm: Origins, Evolution, and Development*, Oxford: Oxford University Press.
Zérah, M.H. (2000) *Water: Unreliable Supply in India*, New Delhi: Manohar.

CHAPTER TWO

"Internetting" Downtown San Francisco: Digital Space Meets Urban Place

Stephen Graham and Simon Guy

The Commercialization of the Internet and the "Sticky" Places of Global Capitalism

In the early twenty-first century the rhetoric surrounding the internet in the 1990s is starting to seem quaintly anachronistic. In a wave of excited hype and speculation, fueled often by the vested interests of corporate media companies, the internet was depicted as supporting the "death of distance" (Cairncross 1997). It was widely portrayed as a fundamentally "anti-spatial" communications medium that somehow "negated geometry" (Mitchell 1996). It was widely predicted that its growth and eventual ubiquity would threaten to undermine the contested materialities of urban life by making everything available, anywhere, and at any time "one click away" (Pascal 1987, see also Graham and Marvin 1996).

Far from causing "territory to disappear," however, it is now very clear that "it is precisely the fact that a multitude of places exist," within the extending and deepening spatial divisions of labor of internationalizing capitalism, that "creates the need for exchange" via (near) real-time communications networks like the internet (Offner 1996: 26). This seems particularly the case now that the internet is becoming commercialized, liberalized and intimately bound up with the digital commodification and delivery of a whole range of corporate and cultural products and services and means of expression (Mosco 1996; Sussman 1997; Herman and Swiss 2000).

As part of this shift the regulated, monopolistic and national communications systems during the Fordist-Keynesian era are being replaced by "unbundled" and highly competitive infrastructure provision regimes. These undermine notions of universal service and the geographically standardized supply of relatively similar services (Graham and Marvin 2001). This means that high capacity broadband networks, for example, are tending to be constructed first in high-demand metropolitan spaces and corridors (Tseng 2000). Consolidating transnational media conglomerates and alliances are laying their own private internet infrastructures, which favor access to their own tie-in e-commerce, digital media and information companies (Schiller 1999). And even the very architecture of the internet is being remodeled as a socially sifting, intensely commodified consumption system. This is based on the construction of "smart" routers, which allow favored users to "bypass" congestion while non-favored users experience "web site not available" signs (Tseng 2000).

As broadband versions of the internet are unevenly constructed, these processes are tending to compound, rather than undermine, the degree to which the privileged users in the large cities that dominate digital innovation, design and application, have infrastructural advantages over other places. Crucially, then, these reconfigurations in the social and economic geometries of digitized, electronic spaces are inseparably bound up with highly contested reconfigurations in the cultures, politics and socio-cultural worlds of the strategic places that dominate digital capitalism: global and "second-tier" cities (Markusen 1999; Schiller 1999). Far from being undermined by the diffusion of digital connectivities, such cities and metropolitan regions are maintaining, and possibly strengthening, their pivotal roles as central arenas of capital accumulation, technological innovation, and financial and economic development (Sassen 1999; Graham and Marvin 2001). This is particularly occurring as such urban spaces, which Ann Markusen (1999) labels the "sticky" spaces of global capitalism, restructure in the wake of the intense geographical clustering of internet-related industries, "dot-com" entrepreneurs, and the service and cultural industries that are designed to meet their needs, within the burgeoning dynamics of urban cultural economies (Scott 1997, 2000).

In such strategic urban sites a tight degree of interaction on the "industrial district" model survives and prospers. In such places flexible, continuous and high value-added innovation continues to require intense face-to-face learning and co-location in (the right) place, over extended periods of time (Zook 2000). This is an irony given that the digital products and services developed through such economic activity can be delivered online to virtually any location (see Veltz 1996; Storper 1997; Markusen 1999). Such strategic spaces of centrality are now driving the production of internet services, web sites and the whole digitization of design, architecture, gaming, CD-ROMs, music, literature, media and corporate services. The cities that are developing such clusters tend to be those with existing strengths in the arts, cultural industries, fashion, publishing, computing and venture capital: New York, San Francisco, Berlin and London to name but four (see Braczyk *et al.* 1999; McGrain 2000; Zook 2000).

This reconfiguration of whole sections of selected central cities as purported "cyber-districts" of intense IT and cultural innovation and commodification inevitably sparks highly contested struggles over the appropriation, occupation and meaning of such urban spaces. Within very short time periods massive influxes of technological and venture capital combine with the in-migration of extremely affluent IT entrepreneurs, and the restaurants, retailers and service industries that target such "high-end" markets. Not surprisingly, such processes spark off major processes of gentrification, speculation and physical and technological reconstruction, which threaten to dramatically alter the cultural, economic and socio-political dynamics of targeted urban sites.

Such trends, however, remain poorly researched. Critical analyses of the ways in which IT is bound up with the socio-political reconfiguration of urban space are made difficult because such agendas tend to fall into the cleavage between social and critical studies of urban neighborhood change and economic and territorial perspectives on the relationships between IT and the restructuring of production. In the former, the massive literatures on gentrification, the reconfiguration of urban public space, and urban social contestation, have largely ignored their relationships with the recent growth of IT clusters in selected urban neighborhoods and central cities (see, for example, Zukin 1995; Smith 1996; Hamel *et al.* 2001). In the latter, the predominance of narrow economic and territorial analyzes on the emergence of such digital "innovative milieu" in reconstructed central city neighborhoods has meant that the socio-political dimensions of such reconfigurations have been largely ignored. As Vincent Mosco puts it:

> there is a great deal of interest in *technopôles* as economic growth engines, some interest in them as new forms of cultural representation, and practically no interest in their political governance, that is, addressing [them] as sites of political power, and their residents as citizens.
>
> (Mosco 1999: 40)

Encouragingly, some recent studies have started to address the socio-politics of the reconstruction of central cities as high-tech innovation sites for technological elites (see Bunnel 2000). In an analysis of the reconstruction of Ann Arbor, Michigan as a strategic corporate technological space, Dolgon (1999) found that the supportive discourses of the local university–corporate–scientific coalition, which portrayed the "rescue" and "revitalization" of declining neighborhoods, managed to obfuscate the processes of displacement, disciplining and exclusion that were its concomitants. Moreover, the reconstruction of selected neighborhoods in the city as chic districts for young professional "digerati" was often portrayed through such supporting discourses as a celebrating of diverse and pluralist community. However, in practice, Dolgon found that such processes tended to reinforce class hierarchies that tended only to include those who could afford the premium costs of accessing new markets for housing and services. Furthermore, the new landscapes created in the process tended to further marginalize those left outside the markets because of unemployment or low incomes.

In an indictment of the even more dramatic reorganization of selected neighborhoods in central San Francisco into "live–work" environments for dot-com entrepreneurs, Solnit and Scwartzenberg (2000) catalogue in detail the displacement, commodification, eviction, real estate speculation, discursive celebration and organized social resistance that surround what they believe is no less than "a crisis of American urbanism." "Something utterly unpredictable has happened to cities," they write:

> They have flourished, with a vengeance, but by ceasing to be cities in the deepest sense. Are they becoming a city-shaped suburb for the affluent? Will the chaotic and diverse form of the city be preserved, but with its content smoothed out, homogenized by wealth?
>
> (2000: 167)

In this chapter we seek to go beyond such analyses by examining the political and spatial contestations surrounding the rapid growth of this gentrifying set of IT-clusters in downtown San Francisco. Our emphasis is on how new, high-capacity internet infrastructures and services, and the technoscientific apparatus to maintain, use and apply such infrastructures, are implicated in the restructuring of politics and landscapes of this particular "global" city. In particular, we focus on the complex urban and technological politics surrounding the "dot-com invasion" of IT entrepreneurs and internet industries into downtown San Francisco. We explore how this urban place has been forcefully appropriated as a strategic site of digital capitalism, under intense resistance and contestation from a wide alliance of social movements struggling to maintain the city as a site of Bohemian counter-culture and social and cultural diversity.

The chapter has three parts. In the first part we set the context by exploring the development politics of San Francisco since the 1970s. Second, we analyze how premium internet infrastructures have been co-produced with new types of built space, to force a radical cultural and economic restructuring of selected districts within the city which have led to a major backlash as a variety of social movements have sought to stop the so-called "Internetting" of the city. Finally, we attempt to draw out implications of this analysis for our understanding of the complex interplay between digital innovation clusters and urban spatiality, in the context of the deepening and stretching of capitalist and neoliberal development processes.

The Context: Reconstructing San Francisco

Unpacking the framing of the debates surrounding the development process in San Francisco necessitates sensitivity to competing discourses around the city's future. In order to identify the socio-spatial implications of development debates and directions, we therefore need to be sensitive to the inter-linking of influential social groups, dynamic development contexts and contested pathways of change in the San Francisco property and real estate market.

Don't "Los Angelize" San Francisco!

Between 1980 and 1986, more than 30 million square feet of new commercial space (the equivalent of 50 Transamerica buildings) were proposed and built in the city. The then mayor Dianne Feinstein called it the "economic salvation" of San Francisco. By contrast, Herb Caen, the late *San Francisco Chronicle* columnist, called it a "vertical earthquake" (both quoted in Redmond 2000). These twin perspectives, or discourses, of pro-growth and growth-control have framed development debates in San Francisco ever since.[1]

Castells's classic study of the establishment of the Mission Coalition Organization (MCO), made up of over 100 grassroots groups and at its height (in 1970–1) totaling some 12,000 people (in a neighborhood of 50,000), who successfully fought the construction of the BART line along Mission Street, highlights the historical nature of urban politics in San Francisco. As Castells argues, "the success of this coalition realized the idea that urban renewal could be stopped and that an alternative pattern of social and urban policies could be developed in opposition to the one inspired by the usually pre-dominant downtown interests" (Castells 1983: 110).

However, Castells notes that the coalition could not survive beyond this single issue, proving too diverse a group to agree on much beyond the need to preserve the community. McGovern also traces the birth of the San Francisco growth-control movement to the late 1960s when plans to evict hundreds of residents in the Western Addition and South of Market (SOMA) area led to demonstrations at City Hall. Critically, McGovern argues that this movement only really gathered pace and sustainability when concerns about the environmental and aesthetic effects of urban development caught the imagination of the urban middle classes. They worried about the environmental consequences of traffic congestion and the way commercial buildings tended to destroy the "Mediterranean feel" of San Francisco (interview quoted in McGovern 1998: 72). Organizing around issues such as the development of the Transamerica pyramid, the battle over the control of urban development that ensued in the 1960s and 1970s was driven less by economic and social issues such as homelessness and working-class dislocation, and more by notions of the aesthetic charm and environmental qualities of San Francisco – especially those that most affected urban professionals living and working in the city core. Walker describes the "environment of civic rebellion" in which this movement was born:

> Pictorial essays touting the splendors of Victorian homes and old skyscrapers cultivated taste for the past . . . local magazines began running articles in rehabbing old homes, and salvage businesses sprang up to save the best bits and pieces of demolished old buildings.
>
> (Walker 1995: 39)

While these new social movements initially only scored small victories around visual amenities, San Francisco nevertheless developed a reputation for its efforts to preserve its unique topographical sense of place. Through a series of innovative planning interventions, often supported by popular and well-organized community action, urban development processes became strongly framed by conservation concerns. Briefly, faced with startling growth patterns through the 1960s and 1970s,[2] concern began to grow about the loss of landmark-quality buildings and the effect of new high-rise office buildings on the visual amenity of downtown San Francisco in particular. Dissatisfied with the limited power of discretionary planning reviews, a number of often community-inspired propositions to limit growth, extract developer contributions and ameliorate the blight of amenity values, appeared through the early 1980s under slogans such as "Don't Los Angelize San Francisco." These efforts culminated, after a series of defeats and new campaigns, in a formal "Downtown Plan" in 1985 and the implementation of a significant cap on new office development termed Proposition "M" in 1986.[3] Thus, for the first time, the once dominant downtown growth coalition became framed by wider public concerns with implications for use, design and location of new property provision (see Loukaitou-Sideris and Banerjee 1998).

With the downturn in development activity from the late 1980s to the mid-1990s that characterized urban development in most developed cities, the effects of Proposition "M" and the Downtown Plan were less keenly felt. However, as McGovern argues, the experience of successfully challenging the pro-growth discourse represented a decisive change in the way San Francisco residents "interpreted downtown development." By the late 1980s affluent middle classes in the city had come to view land use matters "through a progressive cultural prism" (McGovern 1998: 163). Importantly here, the sense of a shared community of "San Franciscans," generated by these protests is key to understanding urban politics as a struggle over identity and the future of the city. Walker notes the almost surreal set of alliances these protests forged:

> the junior league of San Francisco (women from the best families) working hand-in-hand with gay activists (key to the reshaping of architectural taste and rehabilitation of old houses) and alongside African-American neighborhood groups (fighting against a thoroughly racist Black removal strategy of the civic elites).
>
> (Walker 1995: 39)

As we shall see, this shared oppositional sensibility has become a key factor in the contemporary restructuring of urban space in San Francisco.

"Internetting" Downtown San Francisco: The "Dot-Com" Invasion, Premium Internet Systems, and Struggles over the Meaning of the City

> The whole cultural world of San Francisco is being rocked. What is happening right now could affect the whole future of the city. Where we are now is the result of unmitigated development.
>
> (Campbell 2000)

In the year 2000 rental levels in San Francisco exceeded those in New York[4] – the result of a new development boom that swept across San Francisco in the latter half of the 1990s. This new boom was fueled by a massive migration of dot-com entrepreneurs into selected districts of the city core (many of them from the suburban *technopôle* landscape of Silicon Valley, 30 miles to the south). In 2001, developers and investors fell over each other to exploit the few development opportunities arising in a tightly packed urban core to benefit from the "dot-com" boom. Even with the current policy (Proposition "M") in effect, before the current dot-com slump, property analysts Grubb and Ellis predicted the resumption of a very tight development market within 2–3 years.[5] With San Francisco reaching its voter-mandated cap of 950,000 square feet on new office space in spring 2000, and the prospect of multimedia projects totaling at least half a million square feet scheduled to come before city agencies by the end of that year, a new wave of "Manhattanization" is taking place. This raises three related questions. What has fueled this new boom? In what ways is it restructuring socio-spatial relations in the city? And how are the politics of urban development in San Francisco responding?

Dot-coms: The Monster that Ate San Francisco

Since the mid-1990s some of San Francisco's most culturally bohemian, lower- and mixed-income districts, such as SOMA and the Mission, have been the target of an intense wave of investment from dot-com entrepreneurs, internet firms and broadband telecommunications operators, as well as the real estate and service firms geared towards the needs of internet companies. As the high value-added of internet-based economic activity has moved from hardware and software to content, so the "in" districts supporting face-to-face innovation have shifted from post-suburban technopoles, in campus-like environments such as Silicon Valley (see Castells and Hall 1994), to older, "gritty" urban cores that provide the cultural ambience of the "urban frontier" (Smith 1996).

The movement of dot-com entrepreneurs and associated investments north from Silicon Valley into newly constructed multimedia clusters in central San Francisco is perhaps the best example of the reappropriation of the bohemian, urban zeitgeist as a "new urban frontier" to colonize and gentrify for inter-linked complexes of production, domestic living and consumption (see Zukin 1982; Smith 1996). This process has set off spirals of gentrification, attracting considerable investment from restaurants, corporate retailers, property firms, "loft" developers and infrastructure companies, and leading to the exclusion of lower-income groups from the newly "high-end" space (Solnit and Scwartzenberg 2000; see Zukin 1982). As Walker notes:

> A further contradiction of the preservation and gentrification movements is that they assisted in the annihilation of working-class urban culture, pricing most workers out of the urban core. Obscure little South Park (near the foot of the Bay Bridge), once a refuge for a small black residential block, is now a popular eating spot for the denizens of

> Virtual valley, the new hot spot for multimedia electronics and computer magazine publishers.
>
> (Walker 1995: 39)

Rents have exploded and, somewhat ironically for an industry whose products can be sent online anywhere on earth, parking shortages have become critical. As a result, "with cars parked on side walks it's hard to walk around anymore. It takes away from public life" (Solnit 2001). Stories of dislocation abound: "In record numbers, San Francisco landlords are using the state's Ellis Act to evict rents and convert units into [live–work, broadband connected] condominiums" (Curiel 2000). As a result, it is likely that in the next few years "half the arts organizations in San Francisco will lose their leases – and may be out of business" (Nowinski 2000). Lam reports that "Simon, a friend from Hong Kong, is now renting out his walk-in closet for $500" (Lam 2000). The "enemy" here is clear: "E-commerce companies like Dotcomix, Red Ladder and Spinner, the free music company, have arrived and the nearby restaurant caters very specifically for the laptop-bearing newcomers, with smoked salmon filone for breakfast and meeting-places for the nascent companies" (Campbell 2000).

To illustrate the effects of this heating-up of the property market, one non-profit company, Earthjustice (an environmental law firm), leased a central office-space, which was sold for $98 million. This meant that the market rate for office space in the area rose by $80 per square foot, meaning that this charity's rent would have risen from $360,000 to $1.6 million per year (Campbell 2000). Not surprisingly, a move was the only option.

With 35 percent of US venture capital centered on the Bay Area, investments to support cyber-gentrification are quickly restructuring the selected districts of the central city (Solnit 2001). Paul Borsook (1999) outlines the symptoms of what he calls the "Internetting" of the city: commercial real estate rates rose 42 percent between 1997 and 1999; the median-priced apartment was $410,000 by August 1999; median rental for an apartment was over $2,000 per month; and homelessness rates were rising fast. Landlords, backed by the relaxation of rent controls and tenant protection laws by the City Council in the 1990s, have instigated a huge rise in evictions (a 400 percent rise between 1995 and 1997, subsequently running at an official rate of 7.7 per day and an unofficial rate estimated at four times this). In adopting "quality of life" or "zero tolerance" approaches to policing, authorities in San Francisco are also following cities such as New York trying to discipline those who are not tapped into the high-tech, consumerist gentrification process (in this case often the poor, the black and the homeless).

Cyber-gentrification as a "Cultural and Class Purge"

Rebecca Solnit and Susan Scwartzenberg (2000), in their study *Hollow City*, conclude that this broad process is little more than a "cultural and class purge" of the city, invisibly backed up by the intense electronic connectivities of the many competing, globally connected internet fiber networks that are being wired into the old districts. Seventy thousand white collar and high-tech jobs were being created in San Francisco every year in the late 1990s. Homes often were sold for $100,000 over the (already astronomical) asking price. The city authorities "consciously pursued a program of encouraging jobs without addressing the housing issue" (ibid.). And "these newly rich residents spawned a slew of flashy new restaurants, boutiques, and bars that displaced old-economy businesses, especially nonprofits, and the distinct way of life that San Francisco once provided" (ibid.). It has even been suggested that rising stress levels, which have resulted for older residents of gentrifying neighborhoods have been linked with rapid rises in death rates of elderly seniors (Nieves 2000: 12). The result is a severe housing crisis, the expulsion of poorer people from the city (as many can't afford to remain), and accentuating landscapes of social and geographical polarization, as pockets of the city are repackaged as places of work, leisure or living for internet-based businesses and entrepreneurs.

"Ultimate Global Connectors:" Broadband Internet and the Micro-Geographies of "Internet-Ready" Real Estate

Illustrative of this connection between this particular development boom, the internet and social dislocation, is the dispute over so-called live–work space. Within SOMA and the Mission, new types of integrated work and home spaces are being constructed in classic recycled "loft" spaces, created from the refurbishment of industrial era warehouses, factories and office complexes. Within these, broadband internet connections have been closely combined with highly flexible and carefully configured office suites. Labeled "internet-ready" real estate by its inventors, a series of new complexes for interactive media firms are now emerging at the heart of these "cyber districts" (Graham and Marvin 2001).

To their tenants of CD-ROM developers, web companies, digital design consultancies and virtual reality artists, such "internet-ready" real estate offers dazzling suites of global telecommunications connectivity, from up to seven competing companies, direct from the desk, at bandwidths that few other buildings in the world can handle. Emergency power back up, 24-hour security and training, all-important meeting space, secretarial services and advanced fire suppression systems are also provided. The full suite of high-power electrical systems is especially important as "most buildings today are equipped with only 10 percent of the necessary power requirements of an e-commerce or web company" (Bernet 2000).[6]

Contesting the Nature of Live–Work Developments: Constructing Real Estate Projects as Terminals on "Glocal" Digital Networks

To occupying companies, the physical qualities of the chosen buildings (high ceiling height, high-power and back-up electricity supplies) need to be combined with nodal positions on the many privately laid and competitive fiber networks that are the key conduits for internet traffic. "Whose fiber (and what type of fiber for that matter) will be a major consideration in the site selection process. A perfectly built building in the wrong part of town will be a disaster" (Bernet 2000: 17). In a frenzied process of competition to build or refurbish buildings in the right locations, an agent in New York, where similar processes have fueled the explosive growth of "Silicon Alley" south of 41st St., reported recently that "if you're on top of a fiber line, the property is worth double what it might have been" (ibid.).

Growth-control advocates have a very different perspective of such developments, however, and not surprisingly emphasize the wider externality effects of changes of use rather than the role of such buildings as ultra-resilient terminals on premium global-local (or "glocal") electronic networks (see Pawley 1997). Activists working against the dot-com boom argue that lofts rented to commercial tenants are causing further housing problems since some owners are evicting residential tenants in favor of higher-paying office tenants who will pay up to three times the residential rent. Moreover, they claim, these spaces were initially built as housing to help "ease the housing crisis"; losing them to office space or high-end residential use merely serves to accentuate the growing crisis of social polarization afflicting the city. Moreover, every "live–work" that becomes an office creates the need for housing an average of seven more people and removes the "new" unit from the housing stock. Finally, live–works do not pay any normal office impact fees (normally $15 a square foot) for affordable housing, transit, childcare and open space.[7]

Other critiques address the subtle change in the relations between dot-com developments and the public spaces of the street – the traditional site of mixing, spontaneity and serendipity in the city. As the major new dot-com developments in SOMA (some of which now take up entire blocks), become more inward-looking, more geared towards the disciplining of stark boundaries between inside and outside, and more tied to "glocal" electronic connections than local physical ones, so the ways in which they articulate with street frontages changes. "Space inside and outside is monitored and guarded" (Solnit 2001). In effect, this represents a *parallel* process of the privatization of urban space and electronic space, as changes in built form reflect and reinforce the broader shift towards a privatized, splintered and commercialized internet system. To Solnit, these changes "speak to that paranoia of the fear of strangers, of homelessness, of people and crime. It's a meta-message about the way public and private space is changing and being challenged" (ibid.).

Politicizing the Construction of Cyber Districts: Resistance and the Political Backlash

But this struggle is not merely a dispute about rental levels and building uses. Rather, it is a much deeper contestation about the divisive effects of the dot-com boom on neighborhoods and communities. It is a normative struggle over the very idea of what San Francisco as a city actually *is*. It is also an element of a wider social and political struggle against global neoliberalism, the virtualization of urban life and the hegemonic dominance of corporate (network) ideologies and their endlessly repeated celebrations of the unproblematic joys and liberations of IT-mediation (see, Brook and Boal 1995; Mosco 1996; Sussman 1997). Fueled initially by protests over live–work space, broader social and political movements have emerged to challenge the dot-com boom at a deeper level. Echoing the politics of the MCO that Castells studied in the early 1980s, the Mission Anti-Displacement Coalition (MAC), for example, have exposed the fact that dot-com loft developers have "exploited planning loopholes [by arguing that their spaces are live–work spaces], broken zoning ordinances, and neglected to pay millions of dollars in city taxes" (George 2000: 38). Fights have broken out at San Francisco Planning Commission meetings (Kim 2000) and occupations responding to threats of eviction have resulted in mass arrests (Chonin 2000).

In addition, art and cultural activity has become highly politicized. Activist groups have emerged to "defend their ability to survive" in the city despite ever more perilous financial and economic conditions (Solnit 2001). Threatened artist communities have sought to expose and parody the excesses, commercial blandness and neoliberal ideologies of the incoming "dot-comers," with their internet-oriented libertarian philosophy extolling:

> an absolute, Ronald Reagan-esque [ideology with its axioms of] the land of opportunity, everyone doing whatever they want and therefore no one needing help . . . . In the privatized rhetoric of the internet, even the dot-com ads celebrate the idea that you will never need to leave home, and you'll never have to interact with a stranger.
>
> (Solnit 2001)

Finally, political coalitions such as the "Yuppie Eradication Project" are already fighting back, organizing commando-style raids to plaster graffiti and feces on new luxury buildings, and slash the tires of expensive cars (George 2000). Their campaign operates under the banner "The internet killed San Francisco"; one activist argues that "yuppies are moths eating the cultural fabric of the city" (cited in George 2000: 38).

In early 2000, such political coalitions came together to try to force through Proposition "L", a more robust and relevant planning control than the previous Proposition "M". This would impose a moratorium on dot-com and internet-ready development, prohibit live–work loft construction and place an enforced 10 percent levy on remaining projects to subsidize non-profits organizations to stay in the city. In late 2000, Proposition "L" narrowly lost (by 1,315 votes), gaining 141,434 (49.8 percent) votes. An alternative Proposition ("K"), designed to be more development friendly (for instance allowing conversion of lofts to live–work) was overwhelmingly rejected: 171,881, No (60.8 percent), 111,006, Yes (39.2 percent).[8] The battle about Proposition "L" remains vigorous and is likely to have a significant effect on future dot-com development.

Digital Capitalism and the Re-appropriation of Urban Networked Spaces

> Inventive dot-com minds have come up with solutions for many of life's problems, but so far no-one has quite worked out how to preserve the soul of the city at the heart of the boom.
>
> (Campbell 2000)

Clearly, the changing geometries of infrastructural, urban and socio-technical power that we have witnessed in San Francisco are closely bound up with the biased application of new information technologies within neoliberal capitalism and the local politics that frame urban development. Cases such as the "Internetting" of San Francisco demonstrate forcefully that in order to understand such transformations, we need to maintain a parallel perspective. This must address the ways in which urban and technological processes of restructuring are mutually supportive in together reconstructing what we might call "socio-technical geometries of power" made up of strategic urban sites laced together by systems of intense digital connectivities (see Graham 2000).

On the technological side, the active construction of highly capable and customized internet infrastructures for the "sticky" spaces of global capitalism is occurring, while remaining portions of national territories often become neglected or bypassed by such infrastructures. This is part of the shift to a post-national phase of infrastructural development which tends, very broadly, to undermine, or at least challenge, the relatively standardized and equitable infrastructure systems that were constructed in western nations during the Fordist-Keynesian post-war boom (Graham 2000; Offner 2000; Graham and Marvin 2001).

At the same time, on the urban side, in the strategic sites that are emerging as the hotbeds of economic development in the new digital economy, spirals of gentrification and the disciplinary practices of neoliberal and "zero tolerance" urban governance regimes are often tending to squeeze out those without the market or consumptive power to meet the spiraling costs of urban participation. The liberalization of national and transnational regulatory regimes, combined with the commercialization of infrastructure and the highly biased and exclusionary appropriation of selected central urban spaces, means that premium investment in glocal connections are in a sense, becoming "spatially selective" (Jones 1997). Experimental models of urban planning and infrastructure provision are emerging to support the construction of local micro-geographies within strategically significant regions, while withdrawing policies geared towards mass-integration and redistribution (ibid.).

The micro-social and spatial results of this process have been all too clear in contemporary San Francisco. Rather than the whole-scale physical purges of the 1970s and 1980s, more recent urban struggles have revolved around the economic and cultural spheres. Huge leaps in rental values have led to a battle over the cultural identity of long-established districts and communities. Critically, technological and economic might has not had a deterministic impact on San Francisco. As in previous decades, new community groups have sprung up to defend a particular vision of the city with battles fought both on the street and through the courts. While Proposition "L" (the community vision) was narrowly defeated (by less than 1 percent), Proposition "K" (the developers' vision) was overwhelmingly rejected (by over 20 percent). Proposition "K" would have doubled existing growth limits (set by Proposition "M") and its defeat was seen as a major victory by community groups. Beyond the struggle over planning growth limits, wider resistance continues to be evident with the work of groups such as "People Organizing to Demand Environmental & Economic Rights" (PODER). Operating in the Mission district, PODER organizes public protests on a wide range of issues including the "Eviction of long-term neighbors" or "Stopping the Placement of Spanish Conquistador Statues in the Mission District." Moreover, more minor successes in the planning field are evident. For example, in a move to slow gentrification of the Mission District, the Board of Supervisors approved a one-year moratorium in late June 2001 on development of new live–work lofts, tourist hotels, demolition or conversion of housing into commercial space, internet server farms, private business offices and conversions of nonprofit and artist space into commercial buildings.[9] The board also agreed that large commercial projects must receive special permission to build in the neighborhood. The legislation approved by the board mandates the preservation of existing housing in the Mission and would allow new housing projects only if 25 percent of the units are sold or rented below market rate.

Of course, nothing stands still in the global economy. The collapse of the dot-coms and the spillover effect on their suppliers has recently forced many high-tech tenants to abandon or sublet once-prized office space that was in extremely tight supply at the crest of the business boom. In the south Financial District, where many traditional companies had announced plans to expand during the frenzy of the tech bubble, rents have dropped by half. "The tidal wave has gone out. Now the market is getting back to normal" suggests one local commercial real estate analyst:

> A good snapshot of the problem is the 211,000-square-foot Baker & Hamilton building, a renovated warehouse at Seventh and Townsend streets that was totally leased by Organic Inc., a Web services firm, in November. The building today is more than 60 percent empty, according to brokers, and Organic is looking for tenants to take out subleases.
>
> (Levy 2001)

No doubt the economic wheel will turn again before too long and the struggle will recommence. Looking back at the recent past we know that when it does there will be no easy or straightforward accommodation between "digital space" and "urban place." What there will be is the messy complexity that constitutes the sociotechnical restructuring of urban places under capitalism. The legacy of this struggle in late twentieth-century San Francisco recalls the conclusions of Castells in 1983: "What remains from people's efforts is a series of scattered fragments: some programs, many different grassroots groups, a place to live, and the right to keep their identity" (Castells 1983: 171).

Notes

We gratefully acknowledge the support of the Californian Institute for Energy Efficiency (CIEE), which helped to make this research possible.

1 See: www.sfbg.com/News/35/03/03chron.html for a summary chronicle of key events over 50 years.
2 Between 1965 and 1981, office space in San Francisco doubled, reaching a total of 55 million square feet. See Macris and Williams (1999) (available at: www.spur.org/downtown.html).
3 For further background see: Collins *et al.* (1991).
4 See *San Francisco Chronicle* (2000).
5 Ibid. Grubb and Ellis (2001) argue that no change to Proposition "M" will lead to a loosening of the San Francisco office market and will result in 8.6 million square feet scheduled for delivery by 2002. However, within 2–3 years, a resumption of tight office market conditions is likely, as new competitive supply gets delayed by government projects taking up the majority of new development approval allocations.
6 The huge growth in power demand caused by the dot-com revolution in California's economy, has recently exposed the fragility of the electronically powered digital economy to power outages. In this case a rushed liberalization has left the State with reduced power output and electricity resilience at precisely the time when demand was reaching all-time highs, partly because of the growth of the internet, which consumes 8 percent of all US electricity. The results were widespread and extremely damaging outages and a lasting state of emergency (Campbell 2001).
7 See: www.lofts.freeservers.com/.
8 See: www.reproman.com/propm/electionreport.html.
9 See: www.sfgate.com/cgi-bin/article.cgi?file=/chronicle/archive/2001/06/26/MN213793.DTL.

References

Bernet, B. (2000) "Understanding the needs of telecommunications tenants," *Development Magazine*, Spring: 16–18.
Borsook, P. (1999) "How the internet ruined San Francisco," *Salon.com*, News Feature. Online. Available at: www.salon/com/nes/feature/1999/10/28/internet (accessed November 2001).
Braczyk, H.-J., Fuchs, G. and Wolf, H.-G. (eds) (1999) *Multimedia and Regional Economic Restructuring*, London: Routledge.
Brook, J. and Boal, I. (1995) *Resisting the Virtual Life: The Culture and Politics of Information*, San Francisco, CA: City Lights.
Bunnel, T. (2000) "Multimedia utopia? A geographical critique of IT discourse in Malaysia," mimeo.
Cairncross, F. (1997) *The Death of Distance*, London: Penguin.
Campbell, D. (2000) "Dot.com down-side comes home to California," *The Guardian*, 14 August, p. 20.
Campbell, D. (2001) "Blackouts bring gloom to California," *The Guardian*, 19 January, p. 3.
Castells, M. (1983) *The City and the Grassroots: A Cross-Cultural Theory of Urban Social Movements*, London: Edward Arnold.
Castells, M. and Hall, P. (1994) *Technopoles of the World: The Making of 21st Century Industrial Complexes*, London: Routledge.
Chonin, N. (2000) "Art vs space invaders," *The San Francisco Sunday Chronicle*, 3 September, p. 44.
Collins, R., Waters, E. and Dotson, A. (1991) *America's Downtowns: Growth, Politics and Preservation*, London: Wiley.

Curiel, J. (2000) "S.F. rise in condo conversion evictions," *San Francisco Chronicle*, 26 August, p. 1.
Dolgon, C. (1999) "Soulless cities: Ann Arbor, the cutting edge of discipline: post Fordism, postmodernism and the new bourgeoisie," *Antipode*, 31, 3: 276–92.
George, R. (2000) "Mission undesirable," *The Independent on Sunday*, 5 November, pp. 37–9.
Graham, S. (2000) "Constructing premium networked spaces: reflections on infrastructure networks and contemporary urban development," *International Journal of Urban and Regional Research*, 24, 1: 183–200.
Graham, S. and Marvin, S. (1996) *Telecommunications and the City: Electronic Spaces, Urban Places*, London: Routledge.
Graham, S. and Marvin, S. (2001) *Splintering Urbanism: Networked Infrastructures, Technological Mobilities, and the Urban Condition*, London: Routledge.
Grubb and Ellis (2001) "San Francisco office market: Proposition M." Online. Available at: www.grubbellis.com/Corporate/ServiceStrategicStudies.asp?UserGroupID=&LinkID=84&ParentLinkID=0 (accessed January 2001).
Hamel, P., Lustiger-Thaler, H. and Mayer, M. (2001) *Urban Movements in a Globalising World*, London: Routledge.
Herman, A. and Swiss, T. (2000) *The World Wide Web and Contemporary Cultural Theory*, London: Routledge.
Jones, M. (1997) "Spatial selectivity of the state? The regulationist enigma and local struggles over economic governance," *Environment and Planning A*, 29, pp. 831–64.
Kim, R. (2000) "Cops called in to oust rowdy Mission activists at planning Commission meeting," *San Francisco Examiner*, 8 September, p. 1.
Lam, A. (2000) "Busted by Frisco's real estate boom," *San Francisco Examiner*, 5 September, p. A-17.
Levy, D. (2001) "Wide open spaces: dot-com demise sends commercial vacancy rates soaring as rents decline," *San Francisco Chronicle*, 6 May. Online. Available at: www.sfgate.com/cgi-in/article.cgi?file=/chronicle/archive/2001/05/06/RE85752.DTL (accessed November 2001).
Loukaitou-Sideris, A. and Banerjee, T. (1998) *Urban Design Downtown: Poetics and Politics of Form*, Berkeley and Los Angeles, CA: University of California Press.
McGovern, S.J. (1998) *The Politics of Downtown Development: Dynamic Political Cultures in San Francisco and Washington D.C.*, Kentucky, KY: The University Press of Kentucky.
McGrain, S. (2000) "Go to: Berlin," *Wired*, November, pp. 241–56.
Macris, D. and Williams, G. (1999) *San Francisco's Downtown Plan: Landmark Guidelines Shape City's Growth*, San Francisco, CA: SPUR.
Markusen, A. (1999) "Sticky places in slippery space: a typology of industrial districts," in T. Barnes and M. Gertler (eds) *The New Industrial Geography*, London: Routledge, pp. 98–123.
Mitchell, W. (1996) *City of Bits: Space, Place and the Infobahn*, Cambridge, MA: MIT Press.
Mosco, V. (1996) *The Political Economy of Communication*, London: Sage.
Mosco, V. (1999) "Citizenship and the technopoles," in A. Calabrese and J.-C. Burgelman (eds) *Communication, Citizenship and Social Policy*, New York: Rowman and Littlefield, pp. 33–48.
Nieves, E. (2000) "In San Francisco, more live alone, and die alone, too," *New York Times*, 25 June.
Nowinski, A. (2000) "Vanishing point," *San Francisco Bay Guardian*, 12 July. Online. Available at: www.sfbayguardian.com/Aand E/34/41/vanishing.html (accessed November 2001).
Offner, J.-M. (1996) "'Réseau' et 'Large technical System': concepts complémentaires ou concurrent ?", *Flux*, 26 (October–December) pp. 17–30.
Offner, J.-M. (2000) "'Territorial deregulation': local authorities at risk from technical networks," *International Journal of Urban and Regional Research*, 24, 1: 165–82.
Pascal, A. (1987) "The vanishing city," *Urban Studies*, 24: 597–603.

Pawley, M. (1997) *Terminal Architecture*, London: Reaktion Books.
Redmond, T. (2000) "The dot-com road to ruin," *The San Francisco Bay Guardian*, 18 October. Online. Available at: www.sfbg.com/News/35/03/03econ.html (accessed November 2001).
Sassen, S. (1999) "The state and the new geography of power," in A. Calabrese and J.-C. Burgelman (eds) *Communication, Citizenship and Social Policy*, New York: Rowman and Littlefield.
Schiller, D. (1999) *Digital Capitalism: Networking the Global Market System*, Cambridge, MA: MIT Press.
Scott, A. (1997) "The cultural economy of cities," *International Journal of Urban and Regional Research*, 21, 2: 323–38.
Scott, A. (2000) *The Cultural Economy of Cities*, Sage: London.
Smith, N. (1996) *The New Urban Frontier: Gentrification and the Revanchist City*, London: Routledge.
Solnit, R. (2001) "Hollow city," interview in *Feed-Mag*, January.
Solnit, R. and Scwartzenberg, S. (2000) *Hollow City: Gentrification and the Eviction of Urban Culture*, London: Verso.
Storper, M. (1997) *The Regional World: Territorial Development in a Global Economy*, London: Guildford.
Sussman, G. (1997) *Communication, Technology and Politics in the Information Age*, London: Sage.
Tseng, E. (2000) "The geography of cyberspace," mimeo.
Veltz, P. (1996) *Mondialisation, villes et territoires*, Paris: Presses Universitaires de France.
Walker, R. (1995) "Landscape and city life: four ecologies of residence in the San Francisco Bay area," *Ecumene*, 2, 1: 33–57.
Zook, M. (2000) "The web of production: the economic geography of commercial internet content in the United States," *Environment and Planning A*, 32: 422–6.
Zukin, S. (1982) *Loft Living*, Baltimore, MD: Johns Hopkins University Press.
Zukin, S. (1995) *The Cultures of Cities*, Oxford: Blackwell.

CHAPTER THREE

Urban Space and the Development of Networks: A Discussion of the "Splintering Urbanism" Thesis

Olivier Coutard

The publication of *Splintering Urbanism* by Stephen Graham and Simon Marvin (2001) has provided a platform for explorations into the interplay of highly networked infrastructures and the urban societies they support. This chapter discusses some of the main arguments developed in that book, which has been acknowledged as a major contribution to the study of urban networks.[1]

The Splintering Urbanism Thesis

In *Splintering Urbanism*, Stephen Graham and Simon Marvin (2001) develop an exciting analysis of contemporary relations between networked infrastructures and the "urban condition." *Splintering Urbanism* provides a clear argument, accessible even in its more theoretical developments. Like their previous book, *Telecommunications and the City* (1996), *Splintering Urbanism* skillfully combines a descriptive and an analytical/critical perspective in a discussion of the central thesis of the book: the "splintering urbanism" thesis, i.e. the argument that "a parallel set of processes is under way within which infrastructure networks are being 'unbundled' in ways that help sustain the fragmentation of the social and material fabric of cities" (Graham and Marvin 2001: 33).

The contribution of the book to the "splintering urbanism" thesis is threefold. First, the authors heuristically merge different perspectives and different levels of argument in their attempt to "dynamically [work] through social relations in action, rather than [use] essentialized and ossified notions of scale, space, technology, the city, agency, structure or identity" (op. cit., p. 216), to elaborate a "cross-cutting perspective on urban and infrastructural change" (ibid., p. 33) and, ultimately, to study "the parallel and mutually constitutive processes of network unbundling and urban fragmentation" (ibid., p. 215). Their cross-cutting approach is learned (they *do* make use of the 900 entries in the book's reference list), convincing (the argument is coherent and strong) and fruitful (they incorporate a myriad of a priori independent facts and events within one general perspective).

Second, their epistemological position is strong as well as stimulating. They argue that "unbundled configurations of infrastructure networks continuously and subtly recombine with the production of new configurations of urban space; understanding one without the other soon becomes impossible" (ibid., p. 216). And they build a conceptual framework based on this initial assumption (cf. chapter 5 of the book), making use of insights provided by four theoretical approaches: the "large technical systems" approach, "actor network" theory, theories of "changing political economies of capitalist infrastructure" and what they call "relational theories" of contemporary cities (ibid., p. 180).

Third, the splintering thesis is well articulated and well documented. The authors argue that the *unbundling* of infrastructures, reinforced by powerful factors[2] and supported by powerful coalitions of actors, allow for *bypass* strategies, i.e. strategies that seek the connection of "valued" or "powerful" users and places, while at the same time bypassing "non-valued" or "less powerful" users and places (chapter 4). These bypass strategies contribute to the emergence of so-called "premium networked spaces" (ibid., pp. 249 ff.): economic spaces – e.g. foreign direct investment enclaves or business improvement districts – residential spaces – e.g. gated communities – and "social life spaces" – such as commercial malls and theme parks; etc.[3] In particular, *elite* or higher-income groups are increasingly living in places/spaces that are "withdrawn from the wider urban fabric" (ibid., p. 268) in various ways (chapter 6, especially box 6.4, pp. 268–71). This reinforces the "vicious cycle" of splintering, "where attempts at socio-technical secession lead to greater fear of mixing, so increasing pressure for further secession, and so on" (ibid., p. 383). Moreover, the widening gap between connected and unconnected (or disconnected) places and people is all the more worrisome since the world we live in is, increasingly, a *network society* (Castells 1996) in which "the poverty that matters is not so much material poverty but a poverty of connections," which "limits a person or group's ability to extend their influence in time and space" (Graham and Marvin 2001: 288).

Still, solid though it is, the "splintering urbanism" thesis needs to be analyzed further.[4] In this chapter, three of Graham and Marvin's key assumptions or arguments are discussed:

- that, due to the so-called "unbundling" of infrastructure networks, social and spatial disparities in access to network services in the post-monopolistic era are, in general, worse than during the monopolistic era;
- that disparities among spaces in the provision of network infrastructures and services are, in general, socially undesirable;

- that, in general, premium social/economic spaces coincide with "privatized" premium networked spaces.

Note that the ambition of this chapter is not to provide a comprehensive analysis of these questions, but rather to discuss significant cases in a variety of urban and national contexts that appear to contradict the splintering urbanism thesis.

Residential Access to Basic Infrastructure Networks

As Graham and Marvin rightly point out, when discussing issues of "universal service" or "public utilities," we need to be cautious about the possible gap between rhetoric and fact: during the golden age of public utility monopolies (say, from the 1930s to the 1970s), when equal or universal access to basic network infrastructures was the official doctrine, "variations in the quality and degree of social and geographical access to networked infrastructures remained stark" (Graham and Marvin 2001: 185).

At the same time, though, the progressive universalization of infrastructure networks in developed countries cannot be contested. Interestingly enough for our discussion in this chapter, periods of competition within those industries have been concomitant with a higher rate of expansion, as measured by new connections (see Fischer 1992 or Mueller 1997 for a discussion of the telephone sector in the US). But the proper universalization of those services involved massive state support, financially, politically, legally and ideologically (see, for example, Coutard 1997 and 2001 for a discussion of electrification in France and the US). In less than a century, sometimes much less depending on the country and the sector, such networked systems as the telephone, electric light or water supply have become universal.[5] Certainly, a discussion of exactly how universal these basic services are should include the issue of social access and the fact that disparities remain among social groups in terms of access to and, especially, use of these services (Graham and Marvin 1994). But, overall, residential connection to basic network infrastructures gradually became a "social norm" in those countries, and with it the notion that people should not be disconnected *en masse* or for long periods of time.[6]

Thus, in developed countries, *access* to basic network services is not at stake.[7] Admittedly, utility companies are submitted to regulatory and competitive pressures to reduce their costs, and these pressures may fuel changes in their organization, management and attitude to users. The commodification[8] of utility services in many instances leads to practices (e.g. tailored services) and, sometimes, to abuses (excessive prices, forced sales and the like) similar to those observed in more traditional commercial activities. But such commodification has not resulted in vulnerable or "unprofitable" customers being disconnected from basic services. Moreover, the rise in the alternatives to network supplies (from individual wind or solar energy generators to the shift to bottled drinking water) has resulted from new standards and changing attitudes with respect to the environment, health and risks (Barraqué in this volume) and not from deliberate "social dumping" policies by corporatized or privatized utility companies in a liberalized context.

The contrast between industrialized countries and the rest of the world is very significant. In developing countries, domestic access to basic utility services is a reality for only a minority of the population and even in urban areas, access is far from universal. This was the case before the recent "unbundling" period Graham and Marvin refer to and it is still the case today. But the view that the dismantling of public utility monopolies has, in general, led to increased social/spatial disparities in access to network services is questionable. Examples of the development of water supply systems in three cities in the developing world (Windhoek, Namibia; Buenos Aires, Argentina; Delhi, India) suggest that there are many different reasons that may lead to unequal access to basic utility services, and that the reform of such utility services may, to a certain extent, help rather than hamper the expansion of access to these services. Also, a full discussion should include alternatives to networked forms of service provision.

The reason for unequal access to infrastructures may be political, as in the "hydro *apartheid*" regime of water supply in operation in Windhoek until the early 1990s, when the richest 20 percent of the population consumed 60 percent of the water delivered to retail customers, while the poorest half of the population received only 15 percent of the total (Jaglin 1997). The issue of access to water for low-income urban groups was not properly handled in the 1993 reform of the water supply regime. But, as Jaglin notes, reformers did not intend to ignore or negatively impact low-income groups. Rather, there appear to have been three main reasons as to why the reform was focused on urban/rural solidarity instead of on solidarity between rich and poor: the "social debt" of the state to rural areas, the political power of the predominantly rural Ovambo ethnic group and the lack of appropriate statistical instruments (Jaglin 1997: 25; see also Jaglin 1998).[9]

The reason for access inequalities may also be due to insufficient funding of the water supply system. For example, in Buenos Aires, for several decades, water was supplied by a state-owned public enterprise at low prices (flat rates, i.e. independent of volumes consumed) and with high public subsidies, reflecting a hygienist policy of *canilla libre*, or "free-access-to-tap." This policy was the opposite of a liberal (post-monopolistic) policy whereby one would expect water to be supplied as a commercial good, and it also stands in striking contrast to the water policy in Windhoek. What was the outcome of this free-access-to-tap policy? On the one hand, there was indeed free access to tap in the central, richer part of the urban area (6 million people), with very high levels of water consumption per capita, and hence, very large subsidies to those consumers. On the other hand, half a million people were illegally connected to the water network and there was no connection at all for the rest of the area (4 to 5 million people). On top of this, there was very poor quality service and a very high proportion of uncollected bills (Faudry 1999). This situation may be analyzed as deliberate (post) colonial policy; however, there is another way of looking at it: the water supply system did not expand because of insufficient funding and the richer Buenos Aires residents settled where water and other services were readily available. Hence, the slow emergence, under a formally redistributive regime, of what strongly resembles a *premium water-networked space*. Then, in 1993, the service was franchised to Aguas Argentinas, a commercial consortium. The contract stipulated that, during the 30 years of the contract's duration, the new utility should connect the entire population of the franchise area (9 million people) to the water supply network and at least 95 percent of the population to the waste water drainage system. Thirty years certainly is a long time when you are waiting for tap water! But it is not any greater than the time it took to universalize water utility networks in Western European urban areas (not to mention rural ones). The financing of network expansions gave rise to several conflicts (Faudry 1999; and Schneier-Madanes 2001 and Chapter 9 in this book). In particular, customer associations (of already connected customers) took court action to try to avoid having to pay for the expansion of the network – a somewhat discouraging misuse of the "democratic resistance" that Graham and Marvin advocate to counter splintering tendencies (Graham and Marvin 2001: 396). They eventually lost their case against Aguas Argentinas, and the firm was subsequently able to raise the charge. More recently, Aguas Argentinas engaged in ambitious programs of connecting to their network the inhabitants of deprived neighborhoods, including informal settlements (Botton 2004). This constitutes a remarkable case of convergence between social access concerns and the commercial interests of a utility company in the particularly difficult context created by Argentina's economic crisis and the "*pesificacion*" of the economy.[10]

Disparities in water supply may also have technical origins. Take the example of Delhi.[11] In 1995, 60 percent of the 9 million people living in the Delhi region were connected to the publicly owned water supply utility. For 90 percent of the 600 connected households interviewed by Zérah (1997), water was only available either for only part of the day, or at inadequately low pressure, or both. Zérah interestingly shows that quality of service was primarily a function of the distance between the point of supply and the water treatment plant (the longer the distance, the poorer the quality). Second, it depended on the floor on which the dwelling was located. But it was not significantly correlated to the revenue of the household supplied: at similar locations, richer and poorer (connected) households enjoyed a similar quality of service. Zérah also shows that all households affected by irregular (intermittent) water supply engaged in compensatory strategies: storing network-supplied water (two-thirds of households interviewed), pumping underground water (nearly 30 percent of households), reorganizing domestic and other activities (again, 30 percent of households), recycling water, collecting water outside the home, protesting, moving house (1.5 percent of households interviewed declared they had moved because of poor water supply). Importantly, she notes that "low-income households pay [proportionately] more for their water due to compensatory strategies." In a follow up of Zérah's PhD research and a 1997 article, Llorente and Zérah (1998) argue for a reform of the water supply service in Delhi. In particular, they make the important, if debatable, point that individual, domestic, network-based water supply may not be the most appropriate short- and medium-term solution for many cities in developing countries. Rather, they claim that local or national authorities could improve the public good more efficiently by regulating and coordinating the increasing number of competing water suppliers (tank trucks, water jars and bottled water).

These examples suggest the following conclusions. First, the uneven development of, or access to, basic infrastructures in cities in the developing world was in some instances the result of explicit socio-political strategies. In such cities, one can agree that "variations in the modern infrastructural ideal . . . were adapted to the very different contexts of developing cities, where infrastructural configurations were central in structuring power relations between colonized and colonizers" (Graham and Marvin 2001: 88). This was strikingly the case with *apartheid* and post-*apartheid* Southern African countries and cities (cf. Windhoek). However, in other instances, access disparities were the largely unintended (or, at least, not assumed) consequence of formally redistributive rules (Buenos Aires) or the result of technical failure (Delhi). In any case, the remarkable durability of these patterns of unequal (to say the least) access to water and other basic utility services in the cities examined here as in many others, and the fact that such patterns went relatively unchallenged for long periods by the local population, suggest that the "high modern ideal of the ubiquitously networked city" (Graham 2000: 184) may well not have been the ideal in many parts of the world (see Jaglin 2003 for a discussion of this point).

Second, the poorest part of the population, even when it is not deliberately excluded, is hit disproportionately by the uneven access to, and variable quality in network-based water supply. But access of the poorest households to basic services calls for careful design of *ad hoc* technical, commercial and financial measures rather than for standard patterns of network universalization.

Third, the notion of "unbundling" is in many instances misleading as a basis for discussing reforms in many utility services: such services (in particular water) were unbundled in a de facto manner during the monopoly era in developing cities, because a large share of the population of those cities did not have access to networked supplies and relied on a variety of alternative services. Therefore, in those cities, the shift is not from integrated to unbundled infrastructures, but from one pattern of (more or less unbundled) service provision to another.

This discussion prompts observers to be cautious with regard to the rhetoric of network universalization. Cheap-service-for-all policies often end up as bad service for many and no service at all for many more. In many instances, standardized monopolistic networks are, or evolve into, the very premium networks they are supposed to stand in contrast to. Conversely, technical or economic service differentiation may, in specific contexts, prove to be much fairer than standardized schemes of individual in-dwelling network supplies, when such schemes really mean no supply for many in the short or longer run. The example of Buenos Aires shows that the commodification of utility services and the privatization of utility companies may foster rather than hamper improved access to network-based supply, even of the poorest part of the population, within the defined service area.[12] Finally, as the history of industrialized countries suggests, in the long run, the rise and homogenization of living conditions and purchasing power may be the most efficient equalizer of conditions of access to water or other basic utility services. The duration of this process, however, should not be underestimated.

The above discussion was focused on residential access to basic network infrastructures and, more specifically, on access to water, in order to keep the discussion as accurate and concise as possible. Water supply, moreover, is acknowledged as the single most important "service" human beings should have and, of necessity, *do* have access to (even if it is not through a network). It can be claimed that water is very specific among utility industries and that its unparalleled natural monopoly characteristics make it less exposed to unbundling and bypass strategies, and thus to the emergence of premium water networks and "network spaces" (Graham 2002). This would seem to reinforce the argument developed in this chapter: indeed, historical evidence of premium *water*-networked spaces suggests that the distinction between "integrated" and "unbundled" networks is not fully relevant to understanding the interactions between networks and urban dynamics, or at least that the role of network "unbundling" is not as decisive in the production of premium spaces as Graham and Marvin suggest. In some instances, "bundled" networks (in Graham and Marvin's meaning of this term) may contribute to the emergence of premium urban spaces (such as the central, relatively wealthy part of Buenos Aires); while in other instances, the social-political construction of premium urban spaces results in differentiated water supply, rather than the reverse.

Networks, Premium Business Spaces and Spatially Uneven Development

Graham and Marvin also analyze business "enclaves" as another example of infrastructure-reinforced splintering urbanism. They discuss examples such as Malaysia's super infrastructure corridor (2001, pp. 340 ff.), the Singaporian-Indonesian "Sijori growth triangle" (ibid., pp. 346 ff.) or London's "Sohonet" area (ibid., pp. 332 ff.) as examples of "glocal spaces," globally connected to other similar business enclaves throughout the world, while at the same time as loosely connected as possible to their local environments (ibid., box pp. 318 ff.). Those enclaves, they claim, are supported by "intense electronic surveillance, 'fortress' architecture and private policing strategies" (ibid., p. 325). And they emphasize the fiscal logic of these business enclaves in their discussion of business improvement districts (BIDs), "secessionary streetscapes" supported by "a tailor-made form of local government" which allows BID boards:

> to impose property taxes, which are enforced by law, and to use these as an excellent example of "fiscal equivalence" – i.e., all revenues are spent within the district. Free riders, and social or geographical cross subsidies, are thus avoided.
>
> (ibid., p. 261)

Some examples mentioned by Graham and Marvin seem even more striking, such as foreign direct investment enclaves in Brazil, where:

> new auto plants are being equipped at direct municipal and Federal expense with their own private universe of glocal connections . . . [while] at the same time it has been demonstrated that the social provision of basic services is being undermined across cities and municipalities as a whole because of the spiraling public costs of such strategies.
>
> (Graham 2000: 189–90; see also Graham and Marvin 2001: 343–4)

One can agree with Graham and Marvin that tendencies at fiscal secession associated with business districts raise important issues. But Graham and Marvin's argument that money flows generated by international capital hot spots are basically confined to the socio-spatial subsystem formed by these "enclaves," with no "leaks" into their local, regional or national environments, is more questionable. A comprehensive mapping of money flows is a crucial element in the discussion of whether business development areas are subsidized by, or generate revenue in, the broader local community. This raises two questions regarding Graham and Marvin's argument. First, tax revenues represent only a limited share of the total revenue generated by business districts. A generally larger part of this revenue goes towards the remuneration of the people working there and at least part of this money benefits the environment of the hot spots via a trickle-down process. The same holds for private investments within business districts. Second, regarding tax revenues, it must be borne in mind that not all taxes are local. Part of these are collected and their revenue is redistributed at a broader (especially national) level. These two elements should also be integrated into the analysis, as they may change the answer to the (admittedly crude) question of who subsidizes whom.

The *Île-de-France* (Greater Paris), taken as a whole, is an interesting example. Partly due to the French Jacobin (centralizing) tradition, this capital region concentrates a very high proportion (in terms of its area) of the national population, jobs, research centers, public administrations, etc. Advanced telecommunications infrastructures are considerably more developed (in quantity, quality and diversity) here than in any other part of France. Half of total national telecommunications traffic is concentrated within the *Saint-Lazare–Etoile–La Défense* five-mile long "hot spot." Now, does this make the *Île-de-France* as a whole a socially undesirable enclave, a premium network space misappropriating public money? The issue cannot be resolved without taking into account the other side of the coin: every year, due to its high productivity, the *Île-de-France* is the source of a fiscal transfer to the rest of France of approximately 150 billion francs (23 billion euros), which represents 10 percent of the French national budget (Davezies 1999). Hence, the rest of France benefits to a significant extent from the productivity and the competitiveness of the French capital region and its over-concentration of qualified people, jobs, private and public money, and infrastructures. Note that this is not a transfer of part of local tax revenues. It is mainly achieved through social security and income taxes (which are raised and redistributed by the national government).

It is not easy to estimate to what extent this reasoning applies to the economic "enclaves" discussed by Graham and Marvin as money flows resulting from wages and other relevant social arrangements (social protection, health care, pensions, etc.) are not detailed in their study. However, in many of the examples they discuss, it is probable that part of the revenue generated by the enclave flows out into the surrounding environment, if only in the form of the wages of the local workforce. Thus, these examples may not be such clear-cut cases of fiscal equivalence; at the very least this point requires further examination. In any case, this discussion does not boil down to a problem of the differential quality and pricing of infrastructure services inside and outside these areas. Indeed, it is hardly surprising in this respect that "hot spots" are equipped with high-performance telecommunications infrastructures when the neighboring environment is not: the telecommunications needs of both types of space are simply not the same.

But the argument developed by Graham and Marvin is broader. They argue that infrastructural connectivity is progressively replacing spatial contiguity as the key structuring principle of economic spaces, that this process is supported by the development of networked infrastructures and that it exerts pressure on previously existing redistributive fiscal schemes at the local or national level. If this argument is supported by facts, and it probably is, such processes should be regulated on a broader level than the local or national one. In the absence of this broader regulation, most countries or regions feel compelled to compete (including fiscally) in order to attract investors. As the example of the Eastern Paris Eurodisney leisure center shows, this is true in France, a clearly redistributive state, as it is true in states with a more liberal economic system: the decision to locate Eurodisney in the Greater Paris area was obtained thanks to massive "fiscal dumping" and other public subsidies; but the overall result on national income may not be negative. The issue then is not whether the "archipelago" pattern of contemporary capitalism is fairer or harsher to hinterlands than previous patterns of "nationally embedded" capitalism: it *is* harsher (Veltz 1996). Neither is it one of comparing fiscal revenues in the contemporary context with what they would have been if previous fiscal rules had been applied: if such had been the case, keeping to our previous example, there would have been no Eurodisney in Paris and hence no fiscal revenue at all.[13] And neither does the issue consist of comparing the fiscal revenue generated by Eurodisney to the public expenditure incurred. Rather, the issue is to compare the public policies that support the new pattern (by investing in globally connected "hot spots") with their sustainable alternatives – in terms of both the overall income generated and the redistributive effects of such policies, locally, nationally and internationally. However, a thorough discussion of these questions would take us too far from the central argument of this chapter.

Networks and the Specialization of Spaces

The other important issue taken by Graham and Marvin in their discussion of business areas is that such areas involve or depend on the privatization or semi-privatization of a portion of public space. Of course, if all spaces were (*de jure* or de facto) privatized, especially city downtowns, this would indeed constitute a major socio-political issue. However, this issue should be analyzed as a part of a broader process, indeed three interrelated processes: the social specialization of spaces (the decrease of the co-presence of and intercourse between diverse social groups), the functional specialization of space (typically, the spatial separation within urban areas between work, residential, and commercial and leisure spaces) and the development of network patterns of spatial behaviors (i.e. uses of space based more on the existence of connections among – possibly remote – places and less on contiguity or proximity among places).

These processes are acknowledged as fundamental trends in modern urban societies (Dupuy 1992). They are generally associated with the sprawled pattern of suburbanization made possible by the mass diffusion of the automobile. Whether or not this view is historically true, it is sufficient at this point to note that those processes have developed on a large scale *before* the collapse of the "integrated ideal" in infrastructure development and in urban planning more generally.[14]

I would argue that the "privatization" of space results to a considerably larger extent from the social and functional specialization of spaces than from exclusionary urban design or the development of premium networks; that the social specialization of spaces within urban areas generally results from dynamics that have less to do with premium water, energy or even telecommunications supplies than with schooling or safety concerns,[15] for instance; and that the functional specialization of space primarily results from comprehensive forms of urban planning, even though the specialization of space may have paved the way for the subsequent de-integration of the urban fabric. CCTV-surveyed areas and fortress urban design are a consequence of these processes rather than the materialization of new premium networked spaces.

In this discussion, the focus on spatial divisions at the micro level (e.g., between the City of London and the neighboring Hackney borough) is partly misleading as it implicitly rests on the assumption that local, "contiguous" connections are always economically important or socially meaningful. In a spatial organization increasingly characterized by the social and functional specialization of spaces and by networked forms of spatial organization, local relations need not necessarily be important or meaningful. The use of space by individuals or social groups, including less powerful groups, is decreasingly determined by physical distance. As such, one may wonder whether the physical, electronic and private police-enforced locking-up of the city that occurred over the past decade has significantly changed the life of the residents of the neighboring Hackney borough.

More anecdotically perhaps, it can be argued that exclusionary urban design is not always deliberate. Graham and Marvin attribute to "carefully designed local disconnections" the poor accessibility to "inward-looking" shopping malls on foot or by public transit (2001: 5). In support of their argument, it should be noted that locating malls in remote places, which can only be accessed by car, may well prove very efficient as a means of excluding individuals or groups perceived as undesirable or dangerous (mainly teenage male groups on the one hand, and beggars and vagrants on the other). Frequently, however, pedestrians and people traveling by public transport seem to have been ignored (or forgotten) by urban designers rather than deliberately excluded. Consider, for example, the Porte de Bagnolet shopping mall and car-rail hub on the eastern side of the city of Paris. There, the most disadvantaged users are (or at least used to be for a long time, before improvement works were carried out) the automobilists, by far the largest group of customers frequenting the shopping mall. Getting to the mall's car parks from the surrounding road network was incredibly complicated and the paths between those car parks and the shopping center and underground station were desperately ugly and dirty. However, this situation had not been created on purpose: rather than deliberately misconceived, it was just poor architectural design and building management (Margail *et al.* 1996).

More importantly though, improving access to shopping malls by alternative travel modes, desirable as it undoubtedly is, will not solve the general issue of "car dependence" (Dupuy 1999), i.e., the high cost borne by non-car owners in an automobilized society, when, for example, the grocery around the corner closes because of the new shopping center 10 miles away. But on the other hand, even the "locally-disconnected" shopping malls described by Graham and Marvin are accessible to the vast majority of the population in industrialized countries and they cannot be labeled "premium" spaces in the same sense as a gated community in which only households belonging to the richest 1 percent of the population could afford to live.

Conclusion

In this chapter, I have discussed significant processes that in my view stand in contrast to Graham and Marvin's splintering urbanism theory. In closing,

I would like to emphasize the main points of my argument and analyze the limits of this discussion, as well as possible directions for future research.

1 The secular trend in access to basic utility services in old industrialized countries was a trend of universalization. In those countries, water, electricity and telephone networks reached the point of universalization in the course of the twentieth century (see the current debates on "universal services"). In developing countries, the diffusion of networks was limited to parts of urban areas, but not always for deliberate purposes of social exclusion. Moreover, available evidence suggests that regulatory reforms that have affected utility industries worldwide for the past two decades have not systematically aggravated the social disparities in access to basic network services. On the other hand, it is true that universal domestic access to basic utility services in developing countries will not be attained in the short or even in the longer term. Utility services are increasingly differentiated technically and economically, but not always at the expense of aggravated disparities in access among spaces or among social groups, including the lowest income ones. However, the situation of the poorest part of the population remains, in the context of most developing cities, a major policy issue.

2 Not all disparities among spaces in terms of the provision of, access to, and use of, network infrastructures are socially undesirable. Residential spaces and uses at least should be distinguished from business ones. Contrary to what Graham and Marvin seem to imply, in principle, it is not shocking that business districts should benefit from more effective transportation, telecommunications and other infrastructure services than those available in residential districts: the needs of firms are simply different from the needs of households. The key issue is the extent to which the economic achievements of these districts benefit the surrounding population. Social fragmentation among residential areas is more questionable because there is an intrinsic value to the possibility of social relations among diverse groups or individuals (note, however, that co-presence does not mean intercourse).[16] But infrastructure fragmentation is secondary in this process: differential supply of utility services does not make for secessionary spaces; rather, secessionary spaces may allow for differential supply of utility services.

3 The notion of "unbundling" is misleading because it suggests that the provision of basic services was previously "bundled." In cities in developing countries, at least, this has usually not been the case, and the standard pattern of service provision during the "monopoly era" was very much an unbundled pattern, with diverse suppliers and diverse forms of supply. The notion of "premium *network* spaces" (Graham 2000), which implies that, in general, the quality of a given space is determined by the quality of network services supplied to it, is also misleading. The notion of "premium *networked* spaces" seems more appropriate, because it refers more explicitly to social or economic/fiscal

premium spaces supplied with tailored network services. But the specialization of spaces does not result primarily from premium network supplies. In fact, premium network supplies may not even be a good indicator of premium spaces because seemingly homogeneous and standardized infrastructures can coexist with strong socio-spatial segregation in city areas.

Finally, one should be careful not to draw conclusions in relation to processes that clearly have not yet stabilized and that deserve further (primarily empirical) study. I, therefore, fully agree with Graham and Marvin who stress the need to "explore in detail the complex and diverse processes of governance that support and resist processes of splintering urbanism" (2001: 417). Theory-informed empirical studies like that of Graham and Guy (in this volume) are therefore very welcome, because they allow us to explore precisely the role of networks in urban dynamics. The accumulation of such studies will allow us to go beyond the important but frustrating acknowledgement of the diversity of local urban configurations under both "integrated monopoly" and "unbundled supply" regimes. This suggests some specific directions for future research, which might help in understanding the socio-political conditions under which the potentialities of unbundling and bypass are (or are not) actualized, and of the social significance of such processes.

1 The discussion of the "splintering urbanism" thesis would benefit from a more comprehensive discussion of the notion of "splintering" or "fragmentation." Indeed, for analytical purposes, this notion should be distinguished from related notions such as "sprawl" or "residential segregation." On the one hand, as Jaglin (2001) rightly notes, residential segregation may be fully compatible with functional (especially, economic) integration, as evidenced by the *apartheid* urban regime, characterized both by a tightly integrated economic system and sharply segregated (and, indeed, sprawling) cities. Conversely, as noted above, co-presence does not imply intercourse, and the recurring argument that co-presence reinforces social cohesion by making the diversity of society visible would need further critical examination. Conversely, the focus on local regimes of infrastructure provision provides critical insights into fragmentation or integration processes, as they cut across the main layers of local governance (political, institutional, financial/fiscal, etc.).

2 The specific properties of infrastructure networks as large technical systems should be further investigated. Graham and Marvin have rightly favored an intermediary approach between "technological determinism" and "social determinism," which appear equally unsatisfactory as a means for accounting for the interactions between infrastructure development and urban dynamics. Indeed, networks simultaneously shape and are shaped by social-political forces, interests, ideas and institutions. This mutual shaping process affects the spatiality of infrastructure systems. Therefore, the technological characteristics of networks[17] and how they affect the way networks interact with broader urban dynamics deserve particular attention.

3 The specific role of the network *metaphor* should be explored as it seems, in particular, to prompt public authorities to associate universal access to network services with specific social benefits: see the current debate on universal broadband access in Europe. The powerful metaphor of the network and the rhetoric of universal access may well turn out to be, in the long run, a major factor in the development and universalization of networked infrastructures. But this metaphor should also be analyzed and deconstructed with respect to the social normalization to which it gives rise, for example (see Wyatt in this volume), or in terms of sustainable development (see Barraqué in this volume).

Notes

1 This chapter elaborates on two contributions to the *International Journal of Urban and Regional Research* (see Coutard 2002a and 2002b). I would like to thank all participants at the New York City workshop on the Social Sustainability of Technological Networks (18–20 April 2001), and especially Stephen Graham and Ruth Schwartz Cowan, for their comments on an earlier draft of this chapter. I am also grateful to Susan Fainstein, then "debates and developments" editor of *IJURR*, for her helpful editing.

2 The authors list the following: the urban infrastructure "crisis"; changing political economies of urban infrastructure development and governance; neoliberalism and the withdrawal of the state; economic integration, urban competition and the imperatives of global–local connectivity; the development of infrastructural consumerism; the collapse of the comprehensive ideal in urban planning; new urban landscapes; and "new structures of feeling" (chapter 3, pp. 92 ff.).

3 Note that the specialization/functionalization of spaces suggested by this enumeration is itself central to this discussion. I will return to it below.

4 Graham and Marvin's argument is also discussed by Lorrain in this volume.

5 Cf., for example, the eloquent chart on "US households with selected consumer goods 1900–1980," in Fischer (1992: 22), which shows many network goods and services (electricity, the telephone, the car, etc.) that tended to become universal during the twentieth century. A similar chart could be drawn up for all countries in the developed world.

6 See Coutard (2003) for a study of the emerging "right" to networked water supply in France, the UK and Germany.

7 The term access designates here the physical connection to basic utility networks. This is not to deny the existence of stark inequalities in the consumption of basic services and of serious affordability issues.

8 The term commodification refers to the increasing propensity to provide and regulate utility services as mere commercial services, based on notions of full-cost pricing, competition and supplier choice, etc.

9 Although certainly less dramatic, the situation in European countries presents an interesting parallel with that described by Jaglin. Consider, for example, rate averaging for electricity supply and telephone services in France. Historically, the main form of averaging implemented was between urban and rural areas. This was very socially redistributive, subsidizing the predominantly poor farm households. (Incidentally, the massive public subsidies for nationwide rural electrification during the inter-war period were justified by the most powerful lobby in the French Parliament as a counterweight to the enormous losses endured by the French rural population in the trenches during the First World War, another instance of "social debt." (See Nadau (1994: 1209).) But the socially redistributive effect of geographical rate averaging is currently more questionable when in France the poor increasingly live in urban areas, while exurban and rural areas are becoming increasingly populated with upper middle-class households (see

Coutard 1998). Hence, given pricing rules may produce different or even opposite effects when the socio-economic context changes. The UK Parliament failed to acknowledge this change when it required that the newly created regulators of the water supply and other utility industries take account of customers residing in rural areas or customers "of pensionable age," but not of customers on low income. This clause was subsequently changed, however, and low-income customers are now specifically referred to in the regulators' mandates.

10 The water company is in the unpleasant situation of having to pay its creditors and stakeholders in dollars, while its services are paid for in pesos. This currency was massively devalued in 2002 (Botton 2004; Schneier-Madanes in this volume).

11 On Delhi, see also the chapters by Lorrain and Llorente in this volume.

12 The process of differentiation between franchised and non-franchised areas should, however, not be overlooked.

13 Of course, the fiscal argument should not be understood as being absolute: decisions relative to the location of new facilities do not depend solely on local tax rates, and so local taxes do not have to fall into line with the lowest rates worldwide.

14 Note, however, that the specialization at a micro level (within urban areas) combines with metropolization, a major form of de-specialization at the meso level (metropolitan areas taken as a whole) and of polarization at a macro level (e.g. the national level) (Veltz 1996).

15 The extent to which these concerns are based on factual elements will be left as an open question in this chapter.

16 Cf. the research by sociologists and anthropologists on social intercourse in such "public spaces" as malls (see, for example, the special issue on malls in *Flux* 50, October–December 2002, "Paquebots urbains").

17 These technological characteristics include in particular: network externalities (the fact that the quality and diffusion of network-based services are affected by the number of users); the spatiality of network technologies (cable and wireless broadband technologies, for example, can be expected to have very different spatial "behavior"); the temporality of infrastructures (once built, many infrastructures "stay put" for decades or even centuries); and the "momentum" of large technical systems as infrastructures and as organizations (as systems grow, they arguably gain momentum – a combination of weight and velocity, see Hughes 1994 – that renders them less sensitive to external steering and more "autonomous").

References

Botton, S. (2004) "Les 'débranchés' des réseaux urbains d'eau et d'électricité à Buenos Aires: Opportunité commerciale ou risque pour les opérateurs?," *Flux, cahiers scientifiques internationaux Réseaux et Territoires*, 56–7 (April–September): 27–43.

Castells, M. (1996) *The Information Age. Economy, Society and Culture (I): The Rise of the Network Society*, Oxford: Blackwell.

Coutard, O. (1997) "L'organisation industrielle des systèmes électriques sur la longue durée: régularités et variations," in M. Gariépy and M. Marié (eds) *Ces réseaux qui nous gouvernent?*, Paris: L'Harmattan.

—— (1998) "Le 'droit' à l'eau et à l'énergie en France: à propos de quelques évolutions récentes," in N. May, P. Veltz, J. Landrieu and T. Spector (eds) *La Ville éclatée*, éditions de l'Aube, La Tour d'Aigues, France.

—— (2001) "Imaginaire et développement des réseaux techniques: les apports de l'histoire de l'électrification rurale en France et aux Etats-Unis," *Réseaux*, 19(109): 75–94.

—— (2002a) "'Premium network spaces': a comment," *International Journal of Urban and Regional Research*, 26(1): 166–74.

—— (2002b) "Book review of *Splintering Urbanism*," *International Journal of Urban and Regional Research*, 26(4): 867–8.

—— (2003) "La face cachée du service universel: différenciation technique et tarifaire dans le secteur de l'eau en Europe," in G. Schneier-Madanes and B. de Gouvello (eds) *Eaux et réseaux: les défis de la mondialisation*, Paris: La documentation française (Travaux et mémoires de l'Institut des hautes études de l'Amérique latine, University of Paris 3).

Davezies, L. (1999) "Le mythe d'une région spoliatrice," *Pouvoirs locaux*, 40: 38–46.

Dupuy, G. (1992) *L'Urbanisme des réseaux, théories et méthodes*, Paris: Armand Collin.

—— (1999) *La dépendance automobile. Symptômes, analyses, diagnostic, traitements*, Paris: Anthropos-Economica.

Faudry, D. (1999) "La concession des services de l'eau potable et de l'assainissement à Buenos Aires," in D. Lorrain (ed.) *Retour d'expériences, six cas de gestion déléquée à l'étranger*, unpublished report, Paris: Fondation des villes, May.

Fischer, C.S. (1992) *America Calling: A Social History of the Telephone to 1940*, Berkeley, CA: University of California Press.

Flux (2002) special issue "Paquebots urbains," *Flux, cahiers scientifiques internationaux Réseaux et Territoires*, 50 (October–December).

Graham, S. (2000) "Constructing premium network spaces: reflections on infrastructure networks and contemporary urban development," *International Journal of Urban and Regional Research*, 24(1), 183–200.

—— (2002) "On technology, infrastructure, and the contemporary urban condition: a response to Coutard," *International Journal of Urban and Regional Research*, 26(1): 175–82.

—— and S. Marvin (1994) "Cherry picking and social dumping: British utilities in the 1990s," *Utilities Policy*, 4(2): 113–19.

—— and —— (1996) *Telecommunications and the City: Electronic Spaces, Urban Places*, London: Routledge.

—— and —— (2001) *Splintering Urbanism*. London: Routledge.

Hughes, T.P. (1994) "Technological momentum," in Merritt Roe Smith and Leo Marx (eds) *Does Technology Drive History? The Dilemma of Technological Determinism*, Cambridge, MA: MIT Press.

Jaglin, S. (1997) "La commercialisation du service d'eau potable à Windhoek (Namibie): inégalités urbaines et logiques marchandes," *Flux, cahiers scientifiques internationaux Réseaux et Territoires*, 30 (October–December): 16–29.

—— (1998) "Services urbains et cohésion sociale en Afrique australe (Afrique du Sud, Namibie, Zambie): une laborieuse ingénierie," *Flux, cahiers scientifiques internationaux Réseaux et Territoires*, 31/32: 69–82.

—— (2001) "Villes disloquées? Ségrégations et fragmentation urbaine en Afrique australe," *Annales de géographie*, 619 (May–June): 243–65.

—— (2003) *Réseaux et fragmentation urbaine. Services d'eau en Afrique subsaharienne (dossier d'habilitation à diriger des recherches)*, unpublished report, Paris: University of Paris 1.

Llorente, M. and M.-H. Zérah (1998) "La distribution d'eau dans les villes indiennes: quels réseaux pour quels services?," *Flux, cahiers scientifiques internationaux Réseaux et Territoires*, 31/32: 83–9.

Margail, F., G. Doniol-Shaw and P. Legendre d'Anfray (1996) "La gestion du pôle Galliéni – Porte de Bagnolet," *Annales de la recherche urbaine*, 71: 127–36.

Mueller, M.L. (1997) *Universal Service: Competition, Interconnection and Monopoly in the Making of the American Telephone System*, Cambridge, MA: MIT Press.

Nadau, T. (1994) "L'électrification rurale," in M. Lévy-Leboyer and H. Morsel (eds) *Histoire de l'électricité en France, Vol. II (1919–1946)*, Paris: Fayard.

Schneier Madanes, G. (2001) "La construction des catégories du service public dans un pays émergent: les conflits de la concession de l'eau à Buenos Aires," *Flux, cahiers scientifiques internationaux Réseaux et Territoires* 44/45 (April–September): 46–64.

Veltz, P. (1996) *Mondialisation, villes et territoires: l'économie d'archipel*, Paris: Presses universitaires de France (coll. L'Economie en liberté).
Zérah, M.-H. (1997) "Inconstances de la distribution d'eau dans les villes du tiers-monde: le cas de Delhi," *Flux, cahiers scientifiques internationaux Réseaux et Territoires*, 30 (October–December): 5–15.

Part II
Risks, Crises and the Dependence of Cities upon Networks

CHAPTER FOUR

Social Implications of Infrastructure Network Interactions[1]

Rae Zimmerman

Urbanized and soon-to-be urbanized areas are increasingly dependent upon infrastructure transmission and distribution networks for the provision of essential public resources and services for transportation, energy, communications, water supply, and wastewater collection and treatment. In large part, the increasing spread of population settlements at the periphery of cities and the increasing density and vertical expansion of urban cores have increased reliance upon the connectivity that these networks provide. These infrastructure networks are, in turn, not only connected but also often dependent upon one another, at least functionally and spatially, in very complex ways, and that interdependence increases with the addition of new capacity-enhancing infrastructure technologies. The extent of these dependencies appears to be escalating, potentially increasing interactions among the systems and uncertainty in predictions of system reliability and environmental and social effects.

Integrating these services can potentially increase the combined performance of the infrastructures, lower investment costs, and improve urban lifestyles. However, although some spatial and functional coordination of these networks has occurred, the rapid growth in their deployment, advances in network technology, and changes in the distribution of the populations the networks serve, continue to create disruptions in infrastructure systems. These disruptions occur, for example, in the form of street congestion, electric power blackouts, and outages of communications systems to the point where major public services have been disrupted with considerable social implications. These accidents, their disruptions and outcomes are often unpredictable (Perrow 1984, 1999), lending a whole new meaning to the famous old movie title "When Worlds Collide!"

In the US, these interdependencies have, typically, not been studied systematically, given the largely anecdotal information available and the highly dispersed nature of control over the interdependent components. That is changing. The US Department of Energy (US DOE) and the Office of Science and Technology Policy (OSTP), for example, have drawn attention to this issue, underscoring the importance of interdependencies as a critical issue in infrastructure reliability. Their report noted: "The issue of interdependencies among critical infrastructures is a fundamental dimension of critical infrastructure protection. Relative to other infrastructure-specific concerns, infrastructure interdependency has been the least-studied and is probably in the most need of more comprehensive research" (US DOE/OSTP 2001: 11).

The US DOE/OSTP report attributed the previous lack of focus on interdependencies to the interdisciplinary nature of the problem and the lack of integration across various interest groups and stakeholders. As such, the analytical capabilities to evaluate infrastructure interconnections and interdependencies have not been developed to a point where interactions can be easily managed. One reason cited in the report is that: "Although some programs have explored coupling two or three infrastructures, there are no known initiatives developing comprehensive, coupled models of four or more infrastructures" (US DOE/OSTP 2001: 11). Thus, even if the problem is acknowledged, the tools have not been there to address it, though conceptual models began to be introduced at about the time of the US DOE/OSTP report (Rinaldi, Peerenboom, and Kelly 2001).

Interdependencies occur at many different levels within the organization of a system. They occur among components within specific infrastructure facilities, within and among specific infrastructure areas (e.g., energy, transportation), and among those areas and social and environmental systems affecting users of the service and communities that host the facilities. The increasing dependency upon information technologies to manage communications and operations at these interfaces can either exacerbate problems by producing cascading failures or, alternatively, have the advantage of providing greater communication among the systems. Planners, then, can respond to the potentially adverse social and environmental effects of system interconnectivity and interdependencies by relying upon redundancy – a traditional engineering principle – and by building knowledge systems that are based on, and reflect, the infrastructure needs of communities and individual users.

This chapter outlines three key points relating to the issue of the interactions between and among infrastructure networks: interdependencies or interconnectedness and the avoidance of error propagation; redundancy and alternative choices in maintaining operations; and system knowledge that allows for the detection and recognition of threats.

Although these concepts appear to emphasize highly technical aspects of networked infrastructures, they nevertheless have important social ramifications. This is so because technological changes have improved the provision of services of transport, water, electricity, and communications, often transforming the way we live, while at the same time, substantially increasing the fragility and vulnerability of these systems and the service they provide by making them more complex and interdependent (Mitchell 1999).

Types of Interconnectedness and Interdependencies

Interconnectedness refers to a formal linkage between two different systems. A related term, interdependence, connotes a stronger relationship in which two systems not only are connected, but depend upon one another in some way, such as functionally. Not all interconnected systems are interdependent, but all interdependent systems are interconnected. Graham and Marvin citing the work of Gokalp (1992) and Easterling (1999a, b) underscore the inevitable interlinking of networked infrastructures: "Only very rarely do single infrastructure networks develop in isolation from changes in the others" (Graham and Marvin 2001: 30).

Interconnectedness and interdependencies are a natural part of infrastructure design and operation, but as is now well recognized, can be a source of a much wider scale vulnerability or disruption than any single system. "The interdependence of critical infrastructures also enables disruption to propagate" (Schneider 1999: 19). Propagation means that more than one system feels the impact, often in the form of a domino effect.

Hauer and Dagle commenting on the interrelationship of components within power systems note that:

> As a system increases in size, or is interconnected with other systems nearby, it may acquire unexpected or pathological characteristics not found in smaller systems. These characteristics may be intermittent, and they may be further complicated by subtle interactions among control systems or other devices.
>
> (Hauer and Dagle 1999: 22)

As a result of the attack on the World Trade Center, infrastructure was destroyed largely from physical impact on structural elements. However, the cascading effects created by interdependencies magnified the damages within certain systems. For example, the inundation of many infrastructure systems by water contributed to the destructive influences of the physical damage. Water from broken water supply distribution lines and fire fighting was a major contributor to the destruction of parts of the transportation system (i.e., the flooding of train lines) and the electric power systems. This, in turn, disrupted telecommunications by incapacitating Verizon's backup generators, AT&T's telephone switching equipment, and the high-speed internet transmissions provided by nearby "telecom hotels" (Guernsey 2001). As is apparent from the examples below, these cascading effects occur not only as a consequence of terrorism but of accidents and natural hazards as well.

Infrastructure can be interconnected functionally and spatially.[2] Functionally, infrastructure systems can be dependent upon one another operationally, e.g., one system activates the other. Spatially, as infrastructure becomes more dense and compact, and as distributed networks occupy the same conduits in cities, vulnerability to breakages can increase whether or not there is also a functional relationship.

Functional Interconnectedness

Examples of error propagation due to functional interdependency of several different types of infrastructure arise in a number of different ways. Accidents are one way, whether as the result of organizational, cultural, or individual human error, and often affect the same area or type of infrastructure repeatedly. The example of the interdependencies among electric power and other infrastructure provides a rich set of cases. In January 1991 in the northeast United States three New York area airports, including Newark International Airport in New Jersey, ground to a halt when the telephone system failed due to a power disruption (Hevesi 1991: A1). On 9 January 1995, a rupture of high voltage underground electric power lines by construction crews near Newark International Airport in New Jersey again caused the shutdown of the airport (Hanley 1995: A1). The California electric power crisis, when it first began, resulted in the temporary shutting off of the water pumps that serve Los Angeles. Vulnerabilities in the electric power sector can be particularly acute in times of natural disasters, especially where systems are centralized. For example, as Mileti recounts, the high degree of centralization within the electric power grids serving the San Francisco area contributed to the large number of service outages that occurred during the 1994 Northridge earthquake "when 3.1 million customers lost electricity and close to 100,000 homes and businesses were without power for over 24 hours" (Mileti 1999: 59). Thereafter, notes Mileti, the grid gradually became decentralized with greater reliance on renewable systems.

Computers and communication systems can play a major role as initiators of failures in other infrastructure with which they are functionally connected (US GAO 2004). The air traffic control system at Ronkonkoma, Long Island broke down on 7 May 1999 when information technology (IT) hardware and software updates were being made for Y2K compliance for the computers being used by air traffic controllers (Schneider 1999). Many railroad accidents have occurred or have had more serious consequences as a result of a failure of communication systems informing operators of impending problems.[3] The massive blackout in the US and Canada on 14 August 2003 has been related in part to a failure in a software component that, in turn, resulted in the failure of the alarm system to identify the impending electric power problems (Poulsen 2004). The blackout, in turn, led to shutdowns of key transportation and water infrastructures across the affected area. Dependencies and the interconnectedness of electric power with other infrastructure systems occurred prior to the 2003 blackout that contributed to the cascading effect once the blackout occurred.

Wireless communication exemplifies functional conflicts that differ from conflicts that arise with other types of IT. Functionally, different vendors in the US should be connected to provide adequate service to users, but have had incompatible transmission signals. This either produces variations in coverage for users across different parts of the US and the world by different vendors, or a proliferation of wireless towers constructed by each vendor attempting to serve its own customers. Moreover, wireless transmission has produced signal interferences in dense areas. Sanberg (2001), for example, pointed out that wireless devices using radio waves in the 2.4 gigahertz band are blocking each others' transmissions. Wireless transmission can also interfere with electric power transmission and aircraft navigation. This will only get worse given the explosive growth in the number of wireless users (subscribers) and associated cell sites shown in Figure 4.1.

When systems interact or interconnect functionally, the likelihood of failures increases for a number of reasons. One reason is that different systems have been developed at different times, and older systems often have a more difficult time adapting to newer technologies. This is particularly acute where rapidly evolving information technologies are involved. Supervisory Control and Data Acquisition (SCADA) systems, used widely both for the detection of the condition of infrastructure distribution lines and as a basis for operating them, often have to adapt to rapidly changing information technologies before they need to be upgraded. Straayer, for example, points out that: "The

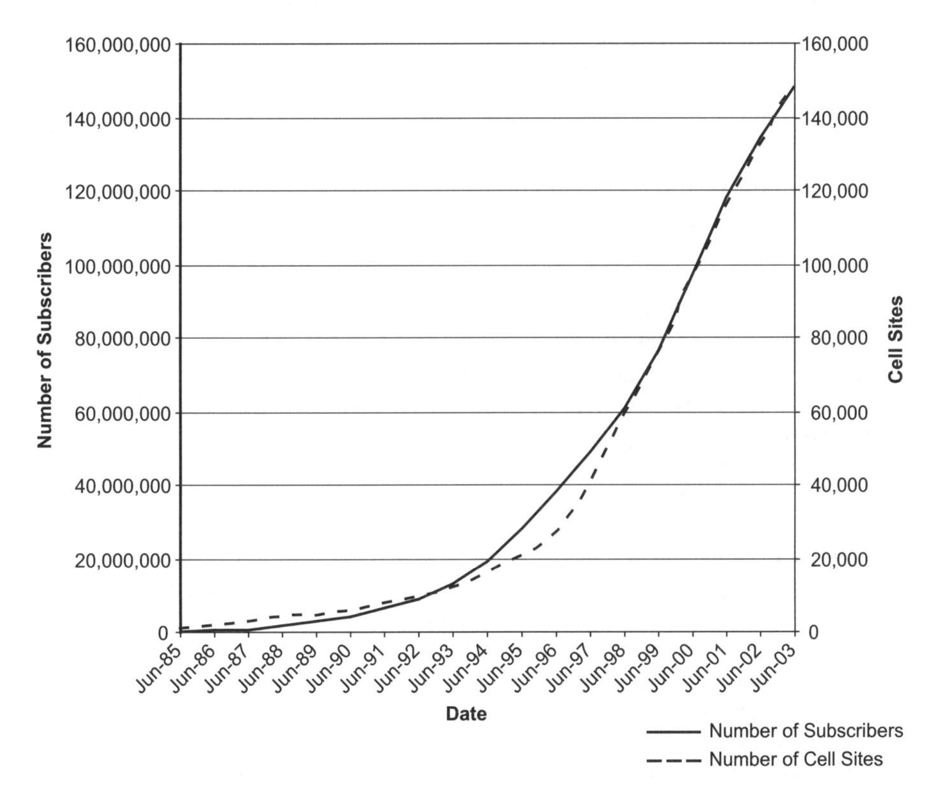

Figure 4.1 *Selected Trends in Wireless Subscriptions and Cell Sites*

Source: Graphed from data provided in Cellular Telecommunications and Internet Association (2004) *Semi-Annual Wireless Industry Survey*, Washington, DC: Cellular Telecommunications and Internet Association. Online. Available at: http://files.ctia.org/pdf/CTIAMidyear2004Survey.pdf (accessed 1 November 2004).

average lifetime for most industrial SCADA equipment is over 10 years, but networking technology changes much faster than that" (Straayer 2000: 26).[4]

Another more fundamental reason for failure of functionally interconnected systems is that activities are not coordinated in time at the same site, coupled with the absence of knowledge of the location of infrastructure facilities and structures. Schneider (1999: 37) has pointed out that backhoe accidents are among the largest cause of failure in telecommunications infrastructure lines. The mapping of infrastructure in order to mark where structures are located is critical to avoiding many of the accidents and ruptures that occur during construction. Although the mapping problem is to some extent technical, it is also political in that different organizations may not want to share this information for reasons of competition and the need to protect their systems from sabotage.[5]

Spatial Interconnectedness

It is largely by design that distribution lines from different utility systems are in close proximity to one another. Historically, utilities took advantage of their common rights-of-way and transportation corridors to locate utility lines.

Most recently, the necessity to co-locate utility lines has been driven in part by the cost imposed by the sheer magnitude of growth in fiber optic cable for telecommunications infrastructure and, to a lesser extent, by the needs of other utility lines. Many estimates exist of the extent to which utility distribution lines are already being installed and projected to be installed. Stix observes that: "Every day installers lay enough new cable to circle the earth three times" (Stix 2001: 81). Iseley and Gokhale (1997) noted that about 150,000 miles of cables and service lines are laid each year in North America – 24 percent will be telecommunications (the smallest size cables though) and between 15–20 percent each of gas, electric cable, water, and sewerage lines account for the rest. Street disturbances account for a substantial amount of the cost of laying underground networks – one contractor indicated that excavating streets accounts for 70 percent of the cost of laying fiber optic cable (Finkelstein 1999). Nunn (1998) estimated that in 1995 the construction of telecommunications distribution lines was 25 percent of non-highway infrastructure. Other estimates are for local areas and are drawn from practical experience. Estimates of the number of utility cuts for Washington, DC, for example, were 5,000 in 1996 and 6,683 in 1998 (Layton 2000a). Underground spaces are becoming crowded and scarce as a result of this demand (Gerwig 2001).

Spatial interconnectedness of infrastructure distribution lines has been increased by placing distribution systems underground. This practice has, in part, been driven by economics, with underground lines reportedly less expensive than overhead lines. It has been more economical to share utility cuts or rights-of-way than to create individual ones for each utility system. The practice of "undergrounding" is very old. In the late nineteenth century, numerous franchises were granted in New York City for laying electrical lines in street cuts, at a cost of a penny per linear foot, which probably led to chaotic street scenes during construction, not unlike what occurs today (Miller 2000).

Installation sites are often in close proximity to user residences and areas in which they conduct their activities. For example, one writer observed that "95 percent of BellSouth's Atlanta customers live within 12,000 feet of fiber-optic lines" (Brister 2000). The increase in the rate and intensity of construction is now well known. The extent to which this is occurring has been so great that some cities are taking drastic measures such as imposing construction moratoria or fees to prevent or restore damage to other infrastructures. The recognition of the negative effects of street openings for utility construction and reconstruction has led to technological and regulatory innovations (Tighe *et al.* 1999).

Innovative Ways of Avoiding Error Propagation from Interconnectedness and Interdependence

Reducing Adverse Interactions through Construction Innovations

Trenchless technologies are one means of avoiding the street disturbances associated with utility line installation. Trenchless technologies are defined as "no-dig" techniques – those that allow direct physical improvements to underground infrastructure systems through innovative construction methods precluding the need for surface excavation (New York University 1999). This technology encompasses a number of different techniques. The use of trenchless technologies has been growing. Back in the mid-1990s, 40 percent of one type of technique – microtunneling – was being done in just one city, Houston (Iseley and Gokhale 1997). Thomson (2000) reviewed the growing extent to which trenchless technology is now used worldwide for utility line installations.

Trenchless technologies are considered to have substantial social benefits as shown in the calculations by Thomson (2000) for the UK and others (Tighe *et al.* 1999). The Transportation Research Board (TRB) report by Iseley and Gokhale (1997) pointed out that an important social benefit of trenchless technology is the reduction in the frequency of intrusions, since lessening the opening up of streets and pavements can extend the life cycle of those structures, reducing the need for reconstruction. The report cites a study showing an increase in pavement life from 10.9 years to 18.5 years in Burlington, Vermont (Iseley and Gokhale 1997: 9). Other benefits include the lessening of noise, dust, and traffic. There are some historic precedents for the social benefits of these techniques. For example, in New York City, the construction of City Tunnel #1, according to historic accounts, was not apparent to the population since the six-year construction partially relied upon vertical shafts located away from the streets (Galusha 1999: 113).

Some of the disadvantages are technical and some are social. Technical disadvantages largely relate to site conditions that do not easily lend themselves to drilling of shafts because of water intrusion, obstructions (including the presence of other utility lines), and land subsidence or settling problems, thereby potentially increasing cost and disruption. Also, maintenance of underground lines could be more difficult and expensive given the greater difficulty of access once they are in place. Social issues that have been identified are that shafts have to go somewhere, and locations for construction facilities have to be selected carefully to avoid neighborhood disruption, which can range anywhere from nuisances to business disruptions around the shaft sites.

Reducing Adverse Interactions Through Regulations

Other responses to street openings pertain to regulatory action. Two examples are briefly discussed below:

Moratoria. Social disturbances from street cutting have occurred to such an extent that Washington, DC imposed a moratorium between 27 March and 8 April 2000 on laying underground fiber, pending a more unified plan from the nine contractors with permits. This is not the first time such a moratorium was imposed in the city. Reactions fall on either side of the issue. On the one hand, the number of times streets are disrupted for laying utility lines due to a lack of coordination results in business losses and social disruption. On the other hand, businesses complain that if construction stops, layoffs will occur, the need of businesses for cable lines will not be met, and the city will lose fees and other benefits from the construction (Layton 2000b).

Permit systems and fees. Different kinds of fee structures are used or are currently under consideration to encourage sound underground construction practices such as damage fees (for damages to other utility lines), access fees, and right-of-way fees (Layton 2000a). Washington, DC has used rental fees, with a pay schedule that is a function of the location of the pipe. Although such fees provide funds for municipalities to restore damaged transportation infrastructure and for traffic management functions in connection with the excavations, they do not directly address the social disruption associated with the frequency and persistence of these activities. Minneapolis also has a fee system, based on installation and repair costs and assessed on a per linear basis. Permit systems, often coupled with fee structures, are becoming a common means of managing network installations. Gerwig (2001) comments on the very large number of such permits, indicating that: "The rule of thumb these days is that building a network requires about one permit per mile from a private landowner or a government body," accounting for about 20 percent of the cost of a new network.

Redundancy and Alternative Choices

A Fundamental Engineering Principle

Redundancy has been a common mechanism to provide alternative services in the event of failures. Redundancy in design and construction is generally considered indispensable and invaluable for structural integrity and reliability of services to the public and, as such, has been a basic and traditional feature of engineered systems.[6]

In their famous work on accidental structural failure, Levy and Salvadori (1992: 55, 56) underscore the critical importance of the concept of redundancy: "In practice, all structural failures may be considered to be due to a lack of redundancy." Specifically, they point out that the advantage of structural redundancy is that: "It allows the loads to be carried in more than one way, i.e., through more than one path through the structure" (Levy and Salvadori 1992: 55, 56). From the perspective of infrastructure planning, redundancies among infrastructure types provide functional flexibility and trade-offs among and between infrastructure systems. Easterling (1999b) points out, for example, that redundant transportation routes between rail, highways, and air routes have afforded the opportunity for specialization as well as intermodal switching for passengers and providers.

The events in lower Manhattan on 11 September 2001 underscore the role of redundancy in restoring the New York region's infrastructure after an immediate shutdown of services (Zimmerman 2003). Although initially, power, information and communication, water, and wastewater infrastructures covering a relatively large area of lower Manhattan were massively disrupted, initial recovery was relatively rapid given the extent of the damage. This occurred because operators could decouple, disengage, and reroute the systems within and around the damaged area. Redundancy initially built into the system created needed capacity. The success of the overall recovery effort for electric power depended on the ability to build new networks and reroute networks to new power centers. New mobile cell towers were brought in by wireless facilities to expand wireless capacity (Guernsey 2001; Young and Solomon 2001). The success of the internet apparently was due to redundancy in the lines – the ability to link high-speed access lines from many different directions.

Failures, however, in some of these systems are attributed to the fact that redundancies may have only been apparent. "Some of those multiple lines travel the same conduits to the same routing centers," and the conduits or routing centers were not redundant enough to withstand the damages (Guernsey 2001).

If many system components are redundant, but a critical link is not, then overall system redundancy and its ability to withstand a system failure can be compromised. This was underscored in the power outage at Newark International Airport and other parts of northeastern New Jersey on 20 June 1997. According to Wald (1997), although power production, transmission, and distribution components were typically highly redundant, substations, such as the one at Public Service Electric & Gas Company's Bayway Switching Station where two networks interconnected, were not redundant. Thus, transformer failures within the substation that served all of the airport feeders caused a system-wide failure (Wald 1997). Recent developments in switching technology using power chip processors may address this limiting factor in the ability of electrical systems to meet greater and more variable demand, with electric power coming from further and further away (Fairley 2001: 41–9).

Redundancy is often reduced when the increased costs of construction to provide for redundancy are weighed against the benefits of added protection given the probability of failure. As failures become more frequent and extreme events become the norm or the consequences more severe, this calculus becomes less convincing. When redundancy is reduced or limited, not only are structures vulnerable but, needless to say, the social systems that depend on them are also at risk. Non-redundancy has led to a number of infrastructure failures (and the potential for others) that are relatively few in number, but when they have occurred, their social impacts were quite catastrophic. The example of water supply (transmission and distribution) will serve to illustrate the importance of redundancy in infrastructure systems that are highly interdependent with their natural and social environments.

Case Example: Water Supply Systems – Redundancy in Service Distribution

There are 2,221 miles (3,575 km) of aqueducts in the US, and many times that number of the smaller distribution lines that extend from them (Scawthorn *et al.* 1991, 2001). According to figures compiled by Herman *et al.*, the total number of distribution lines for about a dozen of the largest cities is many thousands of miles within the official boundaries of those cities alone (Herman *et al.* 1988). The need to meet the growing demand for water supplies further from population centers inevitably increases the extent of these water distribution lines.

Certain configurations of water supply distribution lines that rely upon redundancy introduce greater flexibility than others. Dense urban areas can afford to, and are almost forced to, adopt a matrix structure for their water distribution systems because of the density and the street networks. The New York City system is an illustrative case. The New York City water distribution system, for example, consists of 6,000 miles (9,660 km) of water mains that carry about 1.6 billion gallons per day (6.057 million cubic meters per day) over an area of 308.9 square miles (800.7 square km) (Zimmerman 1999). This amounts to a density of 19.4 miles of water main per square mile (12 km of water main per square km). New York City is, as one would expect, at the higher end of water line density among cities in the US. Elsewhere, the density of water lines, however, shows surprisingly little variation, in general, with the density of cities, ranging from about 16–20 miles of water distribution lines per square mile (9.9–12.4 km per square km), while population per square mile is very variable, e.g., from about 3,000 (Atlanta) to 23,000 (New York) (Herman *et al.* 1988). This seems to suggest that to the extent that density implies redundancy, some cities may promote the flexibility that redundancy offers, but others may not.

New York City has reported the breakage rate for water pipelines as being about 500 to 600 breaks per year. The structure of the water main distribution system contributes substantially to its flexibility in responding to problems. Because of the density and matrix structure of New York City's system of water mains, it can easily switch water from one main to another and isolate damaged areas in the event of a break once the location is discovered, minimizing outages in the short term. One needs to recognize, however, that the key, long-term solution is an ongoing pipe replacement program and condition detection system.

In other areas, such as northeastern New Jersey, for example, the distribution system is linear and branched, with few interconnections among the branches, which is an outcome of the response of infrastructure to more rapid population growth. In such systems, flexibility is far more limited. Once a section of a water line breaks, outages usually occur and last for days. In northeastern New Jersey, breakages are fewer in number than in New York. However, according to one account of the history of recent breaks, effects on water supply can be greater because of the linear (rather than matrix) structure of the water lines.

System Knowledge

Knowledge of infrastructure systems is vital to reducing uncertainty about the nature and effects of infrastructure interactions, but at the same time, knowledge systems can also act as points of vulnerability, depending on how they interact with the infrastructure system. Exclusive dependency upon automated knowledge systems makes infrastructures vulnerable to errors and gaps in these knowledge systems. Computer-based knowledge systems can introduce vulnerability, particularly when they are relied upon to provide knowledge they may not have been designed or programmed to produce.

Designing and Programming Detection Limits

Our technical ability to detect substances in the environment has increased over just the past few decades, and this has had a dramatic effect on environmental policy and management. This experience with chemicals provides an important analogy for infrastructure. Detection is now down to parts per trillion for many chemicals. Computerization aims at expanding that capability in order to avoid catastrophic failures and accidents. However, limitations in the ways in which we use computers have often done the opposite. Two historic examples in the area of chemical detection are instructive. One is the runaway reaction that occurred at the Union Carbide facility at Institute, West Virginia due to the escape of aldicarb oxime, a material used to produce the pesticide aldicarb. This accident occurred shortly after the Bhopal accident at another Union Carbide plant producing aldicarb, where the chemical methyl isocyanate (MIC) had escaped due to water being inadvertently mixed with it during a washing operation. After the Bhopal accident, the Institute plant made use of a computer program to detect gas emissions. The programming included MIC but not aldicarb oxime, so the escape of aldicarb oxime went undetected (Zimmerman 1988). Similarly, the detection of the depletion of the earth's ozone layer went undetected for a long time because the equipment measuring ozone was not programmed to measure low concentrations; that is, the equipment was actually programmed to reject low values considered below the error ranges of analytical models (Benedick 1991). It was not until a ground measurement was made, using conventional technology, that the thinning of the ozone layer was realized. In both cases, we can only notice the risks that we measure.

Although these areas differ in substance from infrastructure, they illustrate the centrality of the design of computer-based knowledge systems in avoiding adverse effects and, thereby, provide lessons for infrastructure. Some examples directly from infrastructure are particularly relevant.

Wastewater treatment. A spill detection system for wastewater flow was installed in San Diego's sewage distribution system, designed only to detect very large spills. As a result, it missed a spill that contaminated Torrey Pines State Beach on 31 July 2001 because the spill was below the detection limit (Rodgers 2001).

Telecommunications. Schneider (1999) cites a number of examples where programming gaps have been identified as contributing factors to numerous outages and disruptions of telephone service and high-speed data networks.

Electric power. As described earlier, one of the most spectacular instances of a computer driven failure was the contribution of a software failure to the massive blackout of 14 August 2003 in the US and Canada (Poulsen 2004).

Transportation. The inability of computer programming to anticipate Y2K touched off high-level investments in readiness efforts. Nevertheless, in Norway, 13 high-speed, long-distance trains failed to start on 31 December 2000 because the computers did not recognize the date (Associated Press 2001).

Adaptability or Flexibility of Detection Technologies

Sensing systems are pervasive throughout utility infrastructures to improve system knowledge and control, but they must be designed into those systems to be compatible with both physical and human systems. Some examples are briefly discussed below (see Zimmerman and Horan 2004 for an extensive discussion).

SCADA, for example, is one commonly used type of system for the detection of conditions within distributed networks of pipelines. Computer chips that are imbedded in fixed structures or materials such as concrete are another example. DNA chips are among the newest additions "imbedded" in fluid materials, e.g. water, that enable the detection of a far greater number of substances in water supplies. Visual techniques for detection are common and include video camera-based techniques with magnified imaging to check road conditions ("Engineering Professor . . ." 2001). Other techniques are used for the detection of currents by photographing bubble patterns as a basis for studies of erosion (Hartman 2001: 12).

The limits of these systems are as important as their capabilities. SCADA systems are subject to interference and not necessarily easily adapted to changes in the information technologies with which they interface to deliver information. Imbedded chips are not easily changed once they are put in place. In an extensive review of diagnostic techniques for sewer systems that employ closed-circuit television, Makar (1999) observed the many advantages and disadvantages of the technique as well as other techniques involving radar and lasers, ultimately concluding that the selection of any given detection technology must be determined by the specific conditions under which it is applied.

Ultimately, this calls for rethinking the management and training of infrastructure engineers and operators. The ability of engineers and operators to manage both the new technologies and the information technology that supports them has become, in many instances, a limiting factor to their application. For example, an Australian survey of engineers' knowledge of information technology was found to be even too limited to effectively understand and develop contracting procedures for others to do the work (Institution of Engineers 2000).

Building a Knowledge Base to Evaluate Interconnected Networks

System evaluation can anticipate adverse interactions to avoid the surprises that occur when different infrastructure systems are combined. The "Trust in Cyberspace" report from the National Research Council observed with respect to information infrastructure that "subsystems spanning distributed networks must be integrated and tested despite their limited visibility and limited control over their operation. Yet, the trend has been for researchers to turn their attention away from such integration and testing questions" (Schneider 1999: 6). There are trade-offs between a reliance on testing v. modeling for system evaluation. With respect to modeling, Hauer and Dagle (1999) imply that when missing information increases as a result of budget cutbacks and increased size and complexity of machinery, greater reliance is placed on modeling. In many industries, relying on modeling for system adjustment is considered a major source of system failure. The complexity that occurs as a result of the dramatic increase in the number of combinations of outcomes that are possible with interactive systems may defy any reasonable ability to conduct direct testing.[7] Yet, some means is needed to evaluate complex systems before they are put into full-scale operation.

Ultimately, an understanding of the complexities of interdependencies among infrastructure systems, and particularly, as they interface with social systems will involve an understanding of how networks function. Barabasi (2002) observed that networks of many different kinds have common properties or operate under certain principles, and those networks that succeed have certain structural properties, such as groupings of nodes or "hubs," that are absent from those that do not succeed. It is this kind of theoretical framework that will need to be applied to infrastructure interdependencies and their connection to social systems.

Conclusions

Every new generation of technological innovation faces a need to connect with the technologies of previous generations. What is different now is the rapid rate and degree of change in the availability of new technologies and the magnitude of the population actually and potentially affected by such changes. Acknowledging system interdependencies, employing redundancy to reduce the level of risk, and building a knowledge base that is consistent with the services these systems perform is likely to require efforts that range from entirely new frameworks or paradigms to innovative practical solutions.

We cannot ignore infrastructure interdependencies and the fact that they can increase uncertainties in what we can expect from our technology. One way infrastructure managers can partially address these uncertainties is to incorporate redundancy and flexibility into the use and design of these facilities. For example, we can compensate for the rigidity of pipes, conduits, tracks, and circuitry by introducing flexibility and redundancy in design, operation, and service for users and host communities.

Constructive approaches to reduce the adverse effects on social and natural environments that occur from interactions among infrastructure systems have been addressed by engineering, management, and planning in a number of different ways. Many recommendations and solutions have been offered that range from relatively small changes in building and design techniques to larger system changes. In concluding this chapter, we will focus on possible organizational and institutional responses to the issues raised by infrastructure interdependencies.

New Ways of Defining Systems for Public Decision Making and Management

Water distribution line failures demonstrate the need for enlarging the management framework from that of a single operating agency with sole responsibility for the condition of the lines to a system of agencies and institutions sharing that responsibility. This new framework makes possible an array of new solutions to problems. Many of these newer potential solutions would target user and impacted populations.

Take the instance of the failure of a water distribution line in a dense urban area. Its immediate technical causes might be freeze-thaw cycles (exacerbated by surrounding water), external chemical corrosion, load stresses created by heavy vehicles, electrical currents from utility lines and subway trains, vibration, undermining of bedding material from improper construction, pressure changes, internal corrosion, and leakage. If one looks more broadly at these immediate causes, many players other than the immediate agency responsible for maintaining the water lines, such as transportation agencies and utilities, should assume some responsibility. These entities often end up experiencing the adverse effects of the breakages as well as contributing to their causes. In large cities with old transit systems, transit agencies often spend large amounts on pumps to remove water from the tunnels, some of which comes from water main leakages and breaks. Potential contributing factors to breakages are vibration and electric currents from trains located near the water lines. Thus, if the conceptualization of water main breakages is enlarged beyond concern over a water main replacement program to the sources of some of the initial causes, a more long-lasting, systems-oriented approach emerges.

New Business Models for Service Integration

Utilities have been using alternative business models for quite some time to coordinate infrastructure to enhance functional integration. These range from modest approaches such as agreements among utilities to share certain functions, to more extensive mergers and buyouts to consolidate conflicting activities. Some of these approaches are discussed below.

Embedded information technologies for communication and control in infrastructure systems. Information technologies are now common in infrastructure industries supporting functions such as surveillance and detection of system conditions using sensors, computerized control of operations and cost accounting, and rapid and real-time provision of information that provides users with increased knowledge of services (Zimmerman and Horan 2004). Linkages have not only been created between information infrastructure and other kinds of infrastructure, but linkages among different kinds of infrastructure have been promoted as well. For example, telephone companies have provided remote-site meter readings for energy and water utilities to avoid the expense to those utilities of having meter readings performed on-site. CellNet, a telecommunications firm, imbeds communications modules into electric meters, enabling them to be read from remote locations (Masud 1999: 22–34).

Bundling of services. Utility "bundling" is a term that refers to multiple services such as communications, electricity, and water being provided by a single entity. Service coordination is the responsibility of the provider, rather than the consumer. There are advantages and disadvantages to this. Advantages are time savings for both users and providers in consolidated services and taking advantage of the utility most familiar with the operation or owning the particular right-of-way for distribution lines. Disadvantages are that consumers of infrastructure services may not want to deal with a single entity for all of their services, preferring to retain their right to select their vendors. Reichman conducted a survey of 1,000 residential users, and the results showed that half of those who switched providers preferred buying individual services, and well over half of affluent college graduate heads of households preferred retaining individual services (Morri 1997). Another disadvantage is the cascading effect that if a failure occurs in a consolidated, bundled system, the user loses more than one service at the same time.

Mergers and buyouts. A more formal and extensive means of coordinating utilities is for one company to become integrated with another organizationally, through mergers or one utility purchasing others in related areas. Large private companies, for example, known for their water supply operations, i.e., Vivendi, Suez Lyonnaise des Eaux, and Thames Water Co., have also been entering telecommunications and energy businesses for some time. However, these consolidations may not always cover or integrate these services in a given geographic area.

Shared rights-of-way (ROWs). Associated with the "undergrounding" of infrastructure and the bundling of services is the use of common rights-of-way for different utility lines (Nunn 1998: 51–72). Although this provides the potential for spatial coordination of distribution networks, it does not necessarily coordinate activities to prevent disruption, for example, from construction, unless all utilities sharing a corridor are held to the same time schedules for installation of lines.

In conclusion, tackling the adverse effects of the interconnectedness of infrastructure systems and turning those interdependencies into positive features will require a greater emphasis upon redundancy, a reliance on new knowledge systems, new means of coordination, and ways of organizing infrastructure businesses that reflect these goals.

Acknowledgment

This chapter is based in part on work supported by several grants from the National Science Foundation: the Institute for Civil Infrastructure Systems (ICIS) Cooperative Agreement No. CMS-9728805, "Integrated Decision Making for Infrastructure Performance," Grant No. 9526057, and "Bringing Information Technology to Infrastructure," Grant No. CMS 0091482. In addition, partial support was obtained from US Department of Homeland Security funding for the Center for Risk and Economic Analysis of Terrorism Events (CREATE). Any opinions, findings, and conclusions or recommendations expressed in this document are those of the author(s) and do not necessarily reflect the views of the National Science Foundation or the US Department of Homeland Security.

Notes

1 This chapter updates and contains portions of "Social Implications of Infrastructure Network Interactions," by R. Zimmerman in the *Journal of Urban Technology*, vol. 8, no. 3 (December 2001), pp. 97–119. Also published in *Flux Cahiers scientifiques internationaux Reseaux et Territoires* (International Scientific Quarterly on Networks and Territories), No. 47, January–March 2002, pp. 54–68. Reproduced by permission.

2 Rinaldi *et al.* (2001) have suggested a four-part typology of infrastructure interdependencies: physical, denoting physical connections between inputs and outputs; cyber, signifying dependency of an infrastructure on information systems; geographic, signifying spatial proximity; and logical, meaning the nature of the relationship between components in different systems. Functional interdependency, as used in this chapter, encompasses physical, cyber, and logical in their schema.

3 These are documented in the extensive reports of the National Transportation Safety Board on railroad accidents.

4 These issues are discussed more fully in the chapters contained in R. Zimmerman and T.A. Horan, *Digital Infrastructures* (London: Routledge 2004).

5 In so far as the ability to map infrastructure has been a technical problem, it is being addressed by newer technologies in the form of geographic positioning systems and

more sophisticated computers to support Geographic Information Systems. A number of cities, such as San Diego, Los Angeles, and New York, have now successfully mapped or are in the process of mapping most or all of their infrastructure. The ability to release that information, however, is constrained by security concerns on the part of municipal officials and owners of the facilities.

6 Although redundancy is generally important, there are instances where it can produce a drain upon resources and increase interdependencies to a point where it creates the very vulnerabilities it is trying to prevent.

7 Testing of complex systems, itself, introduces uncertainty. A number of industrial and infrastructure system failures have occurred during the testing phase, most notably the Three Mile Island accident in Hershey, Pennsylvania in the US.

References

Associated Press (2001) "Millennium Bug Belatedly Bites Norway Trains." Online. Available at: http://www.nandotimes.com (accessed 1 January 2001).

Barabasi, A.-L. (2002) *Linked. The New Science of Networks*. Cambridge, MA: Perseus Publishing.

Benedick, R.E. (1991) "Protecting the Ozone Layer: New Directions in Diplomacy," in J.T. Mathews (ed.) *Preserving the Global Environment*, New York: W.W. Norton.

Brister, K. (2000) "Atlanta's Hungry for High-Tech: City Among Most Wired in Nation." Online. Available at: http://www.accessatlanta.com/partners/ajc/epaper/editions/friday/business_93dde6bec03dc1ae00de.html (accessed 6 October 2000).

Cellular Telecommunications and Internet Association (2004) *Semi-Annual Wireless Industry Survey*, Washington, DC: Cellular Telecommunications and Internet Association. Online. Available at: http://files.ctia.org/pdf/CTIAMidyear2004Survey.pdf (accessed 1 November 2004).

Civil Engineering (2001) "Engineering Professor Designs Pothole Detection System," 36, March.

Easterling, K. (1999a) "Interchange and Container," *Perspecta* 30: 112–21.

—— (1999b) *Organization Space*, Cambridge, MA: MIT Press.

Fairley, P. (2001) "A Smarter Power Grid," *Technology Review* (July–August): 41–9.

Finkelstein, H. (1999) "Revitalizing America's Infrastructure," panel presentation at the *Forbes Magazine* Conference, New York, October.

Galusha, D. (1999) *Liquid Assets*, Fleischmanns, NY: Purple Mountain Press.

Gerwig, K. (2001) "Can They Dig It?" Online. Available at: http://www.teledotcom.com/article/TEL20010319S0026 (accessed 19 March 2001).

Gokalp, I. (1992) "On the Analysis of Large Technical Systems," *Science, Technology, and Human Values* 17: 578–87.

Graham, S. and Marvin, S. (2001) *Splintering Urbanism: Networked Infrastructures, Technological Mobilities, and the Urban Condition*, London: Routledge.

Guernsey, L. (2001) "Keeping the Lifelines Open," *New York Times* (20 September): G1, G6.

Hanley, R. (1995) "Newark Airport is Closed as Crew Cuts Power Lines: New Jersey Widespread Disruption," *New York Times* (10 January): A1.

Hartman, J. (2001) "Remote Sensing Technique Uses Sea Foam to Study Currents," *Civil Engineering* (12 March).

Hauer, J.F. and Dagle, J.E. (1999) "Consortium for Electric Reliability Technology Solutions: Grid of the Future," White paper reviewing recent reliability issues and system events, Richland, WA: Pacific Northwest National Laboratory, December.

Herman, R., Siamak, A., Ardekani, S.G., and Dona, E. (1988) "The Dynamic Characterization of Cities," in J.H. Ausubel and R. Herman (eds) *Cities and Their Vital Systems: Infrastructure Past, Present, and Future*, Washington, DC: National Academy Press.

Hevesi, D. (1991) "Phone Breakdown in NYC Snarls Air Traffic: Long Distance Affected," *New York Times* (18 September): A1.

Institution of Engineers (2000) "Quantifying the Cost and Frequency of Inadequate Information Technology Contracting Practices by Government: A Survey into Government Contracting Practices of Information Technology by the Institution of Engineers, Australia." Online. Available at: http://www.ieaust.org.au/ (accessed November 2000).
Iseley, T. and Gokhale, S.B. (1997) *Trenchless Installation of Conduits Beneath Roadways*, Washington, DC: National Academy Press.
Layton, L. (2000a) "D.C. Official Puts Streets Director on Notice: Control Demanded Over Utility Work," *Washington Post* (16 March): B1.
—— (2000b) "Mayor Vows to Bring Order to Street Work: Longer Moratorium on Trenches Is Possible," *Washington Post* (28 March): B1.
Levy, M. and Salvadori, M. (1992) *Why Buildings Fall Down*, New York: W.W. Norton.
Makar, J.M. (1999) "Diagnostic Techniques for Sewer Systems," *Journal of Infrastructure Systems* 5 (June): 69–78.
Masud, S. (1999) "Utilities Turbocharge Telecommunications," *Business and Management Practices* 33: 24–26.
Mileti, D.S. (1999) *Disasters by Design: A Reassessment of Natural Hazards in the United States*, Washington, DC: Joseph Henry Press.
Miller, B. (2000) *Fat of the Land: Garbage in New York: The Last Two Hundred Years*, New York/London: Four Walls Eight Windows.
Mitchell, W.J. (1999) *E-topia*, Cambridge, MA: MIT Press.
Morri, Aldo (1997) "Bungling on Bundling: Study Shows Customers Want More Choices, Not Less," *Telephony online.* (29 September.) Online. Available at: www.telephonyonline.com/ar/telecom_bungling_bundling_study/ (accessed 5 March 2004).
New York University, Institute for Civil Infrastructure Systems (ICIS) and the Polytechnic University of New York, Urban Utility Center (1999) "Summary Report of the 1999 Life Extension Technologies International Workshop," New York: NYU-Wagner, ICIS.
Nunn, S. (1998) "Public Rights-of-Way, Public Management, and the New Urban Telecommunications Infrastructure," *Public Works Management & Policy* 3: 51–72.
Perrow, C. (1984) *Normal Accidents*, New York: Basic Books.
—— (1999) *Normal Accidents: Living with High-Risk Technologies*, 2nd edn, Princeton, NJ: Princeton University Press.
Poulsen, K. (2004) "Software Bug Contributed to Blackout," SecurityFocus (4 February). Online. Available at: http://www.securityfocus.com/news/8016 (accessed 14 February 2004).
Rinaldi, S.M., Peerenboom, J.P., and Kelly, T.K. (2001) "Identifying, Understanding and Analyzing Critical Infrastructure Interdependencies," in *IEEE Control Systems Magazine* (December): 11–25.
Rodgers, T. (2001) "Torrey Canyon Spill Not on Meter Lines," *San Diego Union Tribune*. Online. Available at: http://www.SignOnSanDiego.com (accessed 2 August 2001).
Sanberg, J. (2001) "Raft of New Wireless Technologies Could Lead to Airwave Gridlock," *The Wall Street Journal* (8 January): B1.
Scawthorn, C., Khater, M., and Rojahn, C. (1991) *Seismic Vulnerability and Impact of Disruptions of Lifelines in the Coterminous United States*, ATC-25 report prepared for FEMA and ATC, FEMA 224, Washington, DC: US Government Printing Office.
—— (2001) "Analysis of National Infrastructure Networks for Seismic Impacts," paper presented at the Workshop on Mitigating the Vulnerability of Critical Infrastructures to Catastrophic Failures, Alexandria, VA.
Schneider, F.B. (ed.) (1999) *Trust in Cyberspace*, Washington, DC: National Academy Press.
Stix, G. (2001) "The Triumph of the Light," *Scientific American* (January): 80–6.
Straayer, R. (2000) "SCADA: The Need for Speed," *Electrical World* 214 (May–June): 26.
Thomson, J.C. (2000) "Worldwide Experience and Best Practices in Mitigating Disruption from Utility Installation," paper presented at the Institute for Civil Infrastructure Systems and Urban Utility Center Second International Life Extension Technologies Forum. New York. August.

Tighe, S., Lee, T., McKim, R., and Haas, R. (1999) "Traffic Delay Cost Savings Associated with Trenchless Technology," *Journal of Infrastructure Systems* (June) 45–51.
US DOE/OSTP (US Department of Energy, Office of Critical Infrastructure Protection and the White House Office of Science and Technology Policy) (2001) *Report*, paper presented at "Infrastructure Interdependencies Research and Development Workshop," McLean, VA, 12–13 June.
US GAO (US General Accounting Office) (2004) "Critical Infrastructure Protection: Challenges and Efforts to Secure Control Systems," GAO–04–354, 25 April.
Wald, M.L. (1997) "Backups Can't Cover for a Substation," *New York Times* (21 June): 26.
Young, S. and Solomon, D. (2001) "Verizon Effectively Rebuilds Network for NYSE," *Wall Street Journal* (18 September): B7.
Zimmerman, R. (1988) "Understanding Industrial Accidents Associated with New Technologies: A Human Resources Management Approach," *Industrial Crisis Quarterly* 2: 229–56.
—— (1999) "The New York Area Water Distribution System: Water Main Breaks – New York City and Northeastern New Jersey Comparison," in *Integrated Decision-Making for Infrastructure Performance*, final report to the National Science Foundation, New York, NY: NYU-Wagner Graduate School of Public Service.
—— (2003) "Public Infrastructure Service Flexibility for Response and Recovery in the September 11th, 2001 Attacks at the World Trade Center," in Natural Hazards Research & Applications Information Center, Public Entity Risk Institute, and Institute for Civil Infrastructure Systems, *Beyond September 11th: An Account of Post-Disaster Research*. Special Publication no. 39. Boulder, CO: University of Colorado.
—— and Horan, T.A. (eds) (2004) *Digital Infrastructures: Enabling Civil and Environmental Systems through Information Technology*, London: Routledge.

CHAPTER FIVE

When Networks are Destabilized: User Innovation and the UK Fuel Crisis

Simon Marvin and Beth Perry

This chapter develops a framework for understanding the potential for innovation among users when a technological network is destabilized. Conventional analyses on network collapses have tended to focus on seeking causes and justifications, assessing reactions and responses and facilitating a rapid return to the "normal" pre-crisis stabilized network. Stepping back from the analysis of cause and effect, however, we argue that network disruption shapes innovative coping strategies that may have the potential for reshaping the user's relations with the network.

This argument stems from an exploratory study of a sample of road users' responses to the UK "fuel crisis" of September 2000. The starting point for the study was the observation that individual car users and employers developed a number of strategies to respond to a situation of limited access to fuel, that seemed unwittingly to be encouraging innovative and, potentially, more sustainable behaviors in car users. The challenge was, therefore, to rapidly capture these altered behaviors, to explore the innovations in more depth and to assess how enduring they might prove to be.

The study examined the strategies, contexts and resilience of new behaviors and innovations in the fuel crisis. This chapter provides an overview of the background to the fuel crisis of 2000, the major responses at the macro-level and implications for our understandings of technological sustainability. It then develops the rationale for the study and presents the methodology and approach that were used. This is followed by an analysis of the study and a presentation of the key results.

A Week When Britain Stood Still

The fuel crisis of September 2000 captured the imagination of the media and public alike. Here was a situation in which small groups of protestors across the country, campaigning against escalating taxes on fuel, succeeded in blockading fuel depots, bringing the distribution network to a standstill. Indeed, such was the success of their action that the government was forced to resort to emergency powers normally reserved for war time.

On the night of Thursday 7 September 2000, over 150 vehicles driven by farmers and hauliers from around the country arrived at the Stanlow oil refinery in Merseyside, north-west England. Earlier in September, crude oil prices had reached $35 a barrel, pushing petrol prices well beyond 80 pence per liter. Protests at the price of oil began to spread from 10 September, largely coordinated by mobile phones. Blockades of oil refineries sprung up around the country as protesters prevented tankers leaving depots without their consent. Refineries across the country closed. As noted by one of the leaders of the protest, the success of the blockade had demonstrated that they could "shake the country to its foundations" (Brynle Williams, *Financial Times*, 15 September 2000). Indeed, over 90 percent of the UK's filling stations ran dry, the emergency services were forced to suspend certain operations, rail services were cut and car users quickly ran out of fuel. Furthermore, from modest beginnings, the fuel crisis spread to a nationwide protest and led to panic buying of fuel and food in the expectation that supplies would quickly run out.

By 11 September, Britain faced widespread shortages of fuel. The government was seriously concerned about the potential impacts on health, other essential services and the economy. The government was granted emergency powers by the Privy Council to regulate fuel supplies, but protestors continued with their blockades and tanker drivers refused to leave their depots. Eventually, the army had to be called in to ensure that essential deliveries of oil could be resumed.

Deliveries of fuel were resumed on Friday 15 September, with an initial focus on the emergency services and essential users. The following week deliveries of fuel to petrol stations for the general public gradually resumed. The hauliers ended the protest for fear of losing public support, but gave the Government a 60-day deadline to reduce fuel tax, with the threat of new blockades. However, despite the organization of a go-slow convoy, this deadline passed with minimal disruption. After a fortnight of chaos and uncertainty, the crisis came to an end.

The aftermath of the crisis focused on recrimination and explanation. A number of causal factors were noted: rising governmental fuel duties, the international price of oil, over-reliance on road transport and over-dependence on fossil fuels (for full analysis of the fuel crisis, see Lyons and Chatterjee 2002). The key question was how a relatively "simple" crisis with a small number of active protestors rapidly escalated into a full-blown national emergency.

A prime explanation can be seen in the vulnerability of technological networks. With relatively little effort, a handful of hauliers were able to destabilize the entire national fuel distribution infrastructure. In part, the new mobile phone and internet technologies were harnessed to coordinate action between otherwise disparate and spatially separated groups. More fundamentally, however, the fuel crisis demonstrated the vulnerability of the distribution network: the just-in-time nature of supply and demand, the fragmentation and loss of control over key systems, the interdependency between networks and dependency of society on the performance of technological networks (see Box 5.1).

Box 5.1 Factors of Network Vulnerability

"Just-in-time Infrastructure"
Infrastructure networks often operate close to their margins as spare capacity has been squeezed out of systems to increase efficiency. When networks are exposed to stress such as high demand the consequence may be system vulnerability. The ability to cope with stress can be reduced as networks operate within reduced margins of error.

"Fragmented Infrastructure"
Network stress can be increasingly difficult to manage as centralized control has been weakened, due to institutional changes – privatization, regulation and contracting out – which have fragmented responsibility for networked infrastructure.

"Network Interdependency"
Increasing levels of interpenetration and interdependence between infrastructures means that when a network fails there are knock-on effects. In the case of the fuel crisis, the destabilization of the fuel distribution network overloaded the public transport network and revealed vulnerabilities therein. (See the discussion by Zimmerman in this volume.)

"Network Dependency"
Society is increasingly dependent on its technical infrastructures and the relations between them so that failures have high consequences and costs. This is particularly so due to the lack of alternatives to reliance on critical systems. Businesses try to ensure high quality reliability but cannot completely transcend these problems.

"Inherited Infrastructure"
The physical infrastructure and emergency management procedures in place in most organizations tend to have been designed for earlier conditions that may no longer be relevant. New arrangements need to be formulated which are flexible enough to deal with and manage change and the inherent unpredictability of events. Network failure brings into sharp focus the difficulty in responding within a framework and network optimized for previous conditions.

Such observations raise key concerns for the sustainability of technological networks and their ability to withstand external shock. From an environmental and public transport perspective, the issue is how more sustainable behaviors and alternative networks can be encouraged that might "spread the load" of travel. It is in this context that the study of car users' responses at the micro-level to the fuel crisis assumes importance.

Studying the Fuel Crisis

We now turn to examine the responses of individual car users and employers to the crisis. While much of the political and policy action focused on ending the dispute, road users and employers were largely left to develop their own strategies for dealing with the disruption caused by the crisis. Road users developed a range of innovations and adaptations to overcome the constraints imposed by the fuel crisis, many of which resonated with the demands of a more sustainable and integrated transport strategy. At the same time, employers, businesses and public sector organizations played a role in creating contexts in which employees and suppliers could consider reshaping their travel patterns.

However, little of the coverage or public debate at the time dealt explicitly with looking at such responses. Given this deficit and in an attempt to capture data that might otherwise be lost, questionnaires were e-mailed to 25 middle managers in Greater Manchester, representing the public, private and voluntary sectors on 13 September 2000. A generic structure for responses was offered (see Box 5.2) in which the managers were asked to keep a diary of their experience during the crisis, focusing on the responses developed by their households and their employers. Sixteen such personal diaries were recorded in written or oral format over the two weeks following the fuel crisis. Users and employer organizations have been made anonymous.

The responses were grouped and analyzed. Three main themes were examined. First, how did a sample of car users cope during the fuel crisis? In particular, what strategies did they develop to deal with the unexpected situation of constrained access to fuel and disruption to both the public and private transport services? Second, what conditions framed the different contexts in which strategies were developed and how did these contexts, in turn, shape the ability of individuals to cope? Finally, to what extent did the alternative travel and behavioral patterns developed during the crisis become embedded in new routines?

The remainder of this chapter will, therefore, be devoted to the range of coping strategies developed, the degree of innovation and adaptation displayed, the organizational responses of work environments and the degree to which respondents expected any changes and new routines to endure.

Strategies for Coping with the Fuel Crisis

Road users were largely left to develop their own solutions to transport within the context of reduced access to fuel. Although the majority of respondents did

Box 5.2 The Diary Questions

- When did you become aware that there was a developing problem with petrol supplies?
- What was your initial response and what action, if any, did you take?
- When did the fuel crisis become more serious for you and how did you respond?
    - Source and amount of petrol – when did you become concerned that you might run out of fuel? Were any supplies available? Did you ever totally run out?
    - Cancelling or postponing trips – did you have to postpone or cancel trips, if so, for what sort of uses – work, shopping, leisure? How did you decide how the car should be used and by whom when there were multiple demands in your household?
    - Using alternatives – were you able to share cars or use public transport as an alternative to your own car? If so, what did you use and what was your experience of the quality and feasibility of continuing to use it?
    - Using your car – did you change the use of the car – how or when you used it? How did car use change over the week?
- How did your workplace respond and what role did you have in any contingency planning?
    - Did work offer any policy or practical guidance regarding work practices and attendance (that you are aware of) during the week?
    - Did your work have a contingency plan and if so what was your role in it?
- How did the crisis change the way you thought about travel options?
- Has the crisis affected how you would consider the use of your car in the future?
- Any other comments you would like to make on the crisis?

not run out of fuel completely, all considered fuel conservation a priority, given that the length of the shortage was unknown. Various innovations were reported, some related to home and others to work routines. They included: trips were cancelled and meetings postponed; days were taken off work; people used alternative transport systems, both public and private, through, for example, car sharing, cycling or walking to and from work. In practice, respondents used a combination of measures, often simultaneously, to overcome the difficulties posed by the crisis:

> I did take measures to conserve the fuel I had . . . I cancelled a social event that I had been involved in organizing and thought hard about making unessential trips in the car . . . I used public transport instead of the car to save petrol. [respondent no. 1]

> I walked into work for the rest of the week . . . I didn't have the flexibility to go and see friends in the evenings . . . I didn't go to the supermarket, I went to the corner shop instead. [2]

> I reduced personal usage of the car and did a bit of walking to local shops . . . I am in the habit of challenging the automatic assumption that a car is best for all business travel. [3]

> The only other change to our family travel patterns was that it enabled my wife to get the kids to walk to school without the usual arguments. She told them we didn't have enough fuel . . . they are not yet old enough to check the truth of her assertion for themselves. [4]

These differing responses can be grouped into three main coping strategies. On the one hand, several respondents did not confront the challenges posed by the fuel crisis directly and put their lives on hold until the crisis was over. This strategy could be described as "suspension." Actions that typify the strategy of suspension include cancellation of meetings or taking days off work. Car users often felt there was nothing that could be done to deal with the crisis and thus waited for a return to normality. On the other hand, a majority of respondents made minor alterations to their usual routines, although the overriding aim continued to be the maintenance of normal patterns of work and home life. Examples of this strategy of "adjustment" include using an alternative transport network to get into work, car sharing or altering domestic routines to conserve fuel. Finally, the third response to the fuel crisis can be classed as "adaptation." This refers to strategies where the normal routine was significantly altered to adapt to the crisis situation, such as working at home.

Despite the fact that the majority of individuals reported the same problems caused by the fuel crisis – getting to work, shopping, family commitments – the strategies developed by users varied greatly. Some were able to implement alternative routines better than others, to adapt, and therefore to cope, more successfully. Others adjusted less well, or suspended planned activities and events until after the crisis. How can we account for these differences in the choice and effectiveness of strategies?

Context and Coping

There is a wide literature on coping strategies which explores the factors inhibiting and enhancing individuals' abilities to cope (Lazarus and Folkman 1984; Carpenter 1992). This shows that the types and nature of coping strategies may differ greatly, and that individuals may employ a variety of different strategies at the same time. Although personal traits and determinants are important factors in affecting the strategies individuals can develop, it is increasingly recognized that contextual resources and support play an important role. This was particularly the case with the fuel crisis. The private car can be seen not only as a component of the transport infrastructure, but as an integral part of a wider socio-technical system that bridges the divide between different personal domains and life spheres.

Given that a breakdown of any part of this network affects all sectors of society, and that individuals have differential access to resources and support, the context for coping assumes increased importance. This pilot study therefore examines the contexts in which individuals coped and the ways in which these contexts may have shaped or framed the ability of individuals to respond to the crisis. The responses from the questionnaires revealed that work-based travel was most frequently prioritized over home-based travel. Indeed, few alternative strategies were reported in relation to personal activities. For this reason, the focus here will be on the work context, in particular, the responses of employers to the fuel crisis.

A range of employer responses were reported, ranging from employers demanding that employees make their way into work under any circumstance, to providing alternative transportation for employees themselves (see Table 5.1). In a few cases, employers even changed their working patterns, allowing and enabling teleworking from home.

Table 5.1 *Employer Responses to the Fuel Crisis*

Strategy	*Level of flexibility and support offered*
Staff expected to get into work no matter what	Low
Information distribution on availability of supplies	Low
Response on an individual basis	Moderate
Cancelling/rescheduling meetings	Moderate
"Encouragement" to car share/find alternative means of transport	Moderate
Workers allowed to work from home	High
Organizing car pools	High
Providing alternative transportation	High
Contingency planning to get staff into work	High
Teleworking and home conferencing over the internet in real time	High

These reactions to the fuel crisis can be grouped into two main strategies that differ from each other in terms of the degree of flexibility and support offered to employees. Employer strategies can be seen to be either *passive* or *active*. On the one hand, the strategy of insisting that staff get into work or distributing information about the state of the crisis is passive and marks out a non- or minimally supportive context for innovation. On the other hand, providing alternative transportation, contingency planning and facilitating home working can be seen as hallmarks of an active response to crisis, through creating a supportive context in which individual coping strategies can be developed. On the whole, employer responses in this sample demonstrated a higher level of flexibility and innovation than individual responses.

Clearly, not only individuals, but also employer organizations, had to develop strategies to cope with the unusual demands placed upon them by the fuel crisis. However, the data gathered reveal that some road users were better able to innovate than others. Similarly, analysis shows that employer responses varied in the level of support offered to employees. The challenge is to see how these two results are linked.

Figure 5.1 represents a grid typology with four main sections, designed to represent the relationship between a supportive context and levels of coping and adaptation. Each person surveyed was grouped according to their ability to adapt and the level of support offered to them by their work environment. Users and employers adopted a variety of different coping strategies and were therefore grouped according to their overall ability to adapt and innovate. Four groups of people emerge, using different coping strategies and being supported to a greater or lesser degree by work environments. Each group would seem to share key characteristics in terms of how they coped and the language they used.

Group A: High Ability to Adapt/Highly Supportive Context

This group of users took advantage of the opportunities afforded to them and were keen to innovate. They were well supported by their work environments, which allowed a variety of different responses to be employed to cope successfully with the crisis:

> *Suzanne, an employee of a major telecommunications company, was not able to get to work due to the fact she lived in the country, and further needed to conserve petrol for critical home use. However, having access to the company intranet meant that she was able to access all relevant information needed to work from home and even attend a training session in real time using interactive services.*

Group B: Low Ability to Adapt/Non-supportive Context

A second group of people had to cope alone. This group of people was only minimally supported by their respective work contexts and was consequently

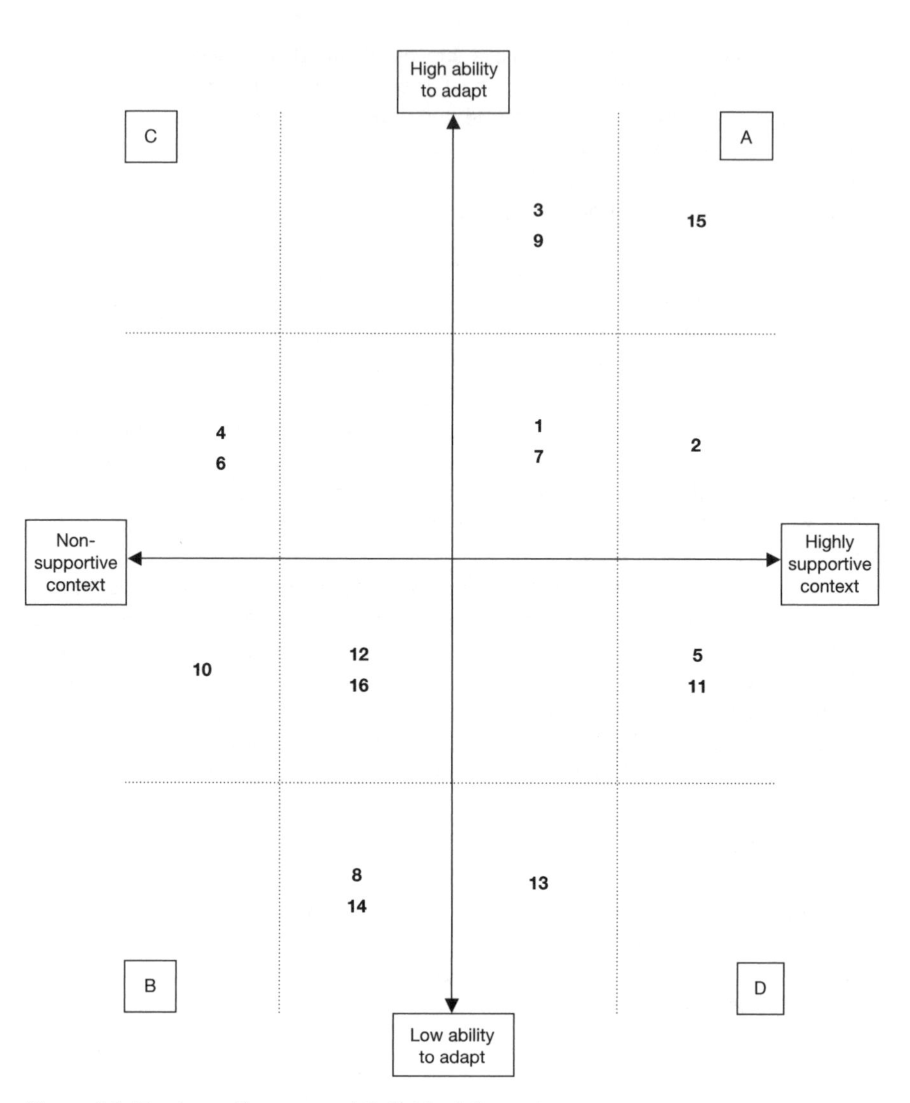

Figure 5.1 *Employer Contexts and Individual Strategies*

left with little ability to innovate. The comments of these respondents indicated a frustration at not being able to adapt:

> *Nigel, a manager for a large factory, could not consider alternatives for getting into work, and his employers introduced no contingency planning, partly due to the fact that the workplace was only minimally affected.*

Group C: High Ability to Adapt/Non-supportive Context

A third group coped well in a relatively unsupportive environment. This group of self-helpers managed to adjust by drawing on other personal or infrastructural resources:

> *Alison, an employee of a pressure group, walked into work due to the proximity of her home to her workplace and recounted how people were left to arrange things between themselves to cope with the crisis.*

Group D: Low Ability to Adapt/Highly Supportive Context

In contrast to these self-helpers, a final group of respondents were well supported by their work environment, with many opportunities provided to facilitate change and adaptation to the crisis situation, yet did not need or want to take advantage of this. Various reasons account for this failure, including attitudinal barriers, for example:

> *David, a member of the police force, believed that it would be impossible for him to get into work on public transport at adequate times and therefore did not try alternatives.*

Approximately two-thirds of the sample falls into the first two groupings. Those receiving a medium or high level of support from their respective work places were able to better innovate than those with low levels of support. Indeed, certain user responses were only made possible with the help or assistance provided by employers, such as getting to work via alternative transportation or working from home. Often employers would establish car-sharing pools over the intranet to facilitate employees getting into work. In such cases, context plays a supportive role, enabling effective coping. On the other hand, context has also been shown to hinder the ability of an individual to cope, if for example an employer remains inflexible in the face of disruption and change. At the same time, however, it should be noted that context is only one determinant of coping strategies and effectiveness. Other factors such as personal support, willingness to take advantage of opportunities and attitude may also have played a role and there are several examples where context has been irrelevant to an individual's ability to adapt to a crisis situation.

On the whole, however, grouping road users via the grid-typology above would seem to indicate that there is some link between social context and ability to cope in a crisis – in this respect, coping is shaped and framed by that social context. However, this is only an exploratory analysis and further research could be done to quantify or flesh out the exact nature of the relationship between context and coping and what this means for interventions designed to encourage transference from one network to another and for encouraging more sustainable behaviors.

The Resilience of Adapted Behaviors

Another key indicator of sustainability relates to the resilience of new behaviors and innovations once they have been introduced. The fuel crisis brought society's reliance on a car-based economy into sharp focus, at a time when transport policy is attempting to shift users away from the car towards more sustainable forms of transport. In addition to impacting on economic and environmental policy, the crisis provided an unexpected, and almost unprecedented, opportunity to examine the difficulties and possibilities of re-orientating the economy and lifestyles away from reliance on the private car.

The importance of the adaptations and innovations adopted during the fuel crisis lies, therefore, in the potential for long-term change towards more sustainable transport behaviors. Many of the changes and alternatives reported would reduce pollution and congestion and lead to healthier and safer urban areas. Some changes may be more desirable than others; for example, working at home may not be a viable widespread alternative to reduce car usage, while the continued use of car pools, collective transport options, driving more slowly to conserve fuel and reducing work-based travel through teleworking and video conferencing could all be integrated into sustainable infrastructure policies. Other non-work related changes could be encouraged in the interests of sustainability: walking, cycling, and using alternative transport networks would all contribute to stemming the environmental problems associated with excessive car use.

Indeed, much of contemporary policy development has focused on sending signals to road users to "think about your travel," in order to create a context for increased use of public transport and other more sustainable transport behaviors. At the same time, major employers have been encouraged to develop travel plans to manage trip generation more effectively, to create incentives for car sharing and public transport use, and to manage demand more effectively. The fuel crisis created a context where users and employers had to consider such changes rapidly and urgently in order to cope with the constraints imposed by fuel shortage. There was therefore a possibility that the unanticipated experience of having to cope without a car might have facilitated the adoption of more long-term sustainable transport behaviors.

With this in mind, questionnaire respondents were asked whether the fuel crisis had changed their attitude towards the car and whether they thought the changes and innovations introduced would be resilient. The results are shown in Table 5.2. With only three exceptions (nos. 2, 10, 16), users anticipated that their usual routines and future use of the car would not be affected by the fuel crisis. Furthermore, this would seem to be irrespective of individuals' ability to cope during the crisis or the level of contextual support they received. In this respect, the fuel crisis can be seen as a missed opportunity to re-orient travel behavior. Three elements at least should be mentioned when seeking to account for this missed opportunity.

First, the fuel crisis lasted approximately one week and Britain rapidly returned to the normal pre-crisis situation. Because the crisis was ultimately

Table 5.2 *The Anticipated Resilience of Change*

Grid section	*No.*	*Comments on likely resilience of changes*
A	1	I'll still use my car in the same way
	2	I have decided to cycle more for leisure traveling. I am also more conscious of the impact speed has on fuel consumption
	3	The car is the only possible way of getting to work . . . public transport has an awful long way to go before becoming a viable alternative
	7	The crisis didn't change the way I thought about travel options – the only change we discussed was keeping at least half a tank of fuel in each car as a contingency
	9	I certainly won't consider commuting by public transport when it could take me twice as long as in the car
	15	I would like to say that I will change the use of my car, but in reality, living where I do means that it is already a costly business and the whole family tries to rationalize trips as much as possible
B	8	[No comment]
	10	I believe I will be able to persuade my employer to pay for conversion of my next car to LPG or any other hybrid
	12	I haven't had time to think about it
	14	My car is essential for my job
	16	The crisis confirms my view about the importance of public transport . . . we tend to use the car in a sensible and careful way and avoid making unnecessary trips anyway
C	4	My use of the car in the future will probably not change much
	6	I think harder about non-essential use of the car, but I imagine that this will wear off sooner rather than later
D	5	The reality is that for my work it is impossible to get into work at adequate times
	11	The crisis has affected how I think about the car surprisingly little
	13	I haven't changed the way I will use the car in the future. I need it for work and so I will carry on as before

temporary and short term it meant that there was insufficient time for new behaviors to become embedded and routinized.

Second, the interview responses revealed that a poor experience of public transport in terms of comfort or journey times meant that the car would remain, for the time being at least, the preferred mode of travel. Habit and ease were also powerful factors in the continued use of the private car, even in the cases of respondents who recognized that they could viably make greater use of alternative modes of transportation. Even where respondents recognized the pressing need to reduce reliance on the car as a mode of transport, this was rarely seen as an action they would personally undertake.

Third, a key factor in the lack of changed behaviors following the crisis was the post-crisis context of support. Indeed, the coping literature suggests that any innovations or behavioral changes adopted during crises, whether of a personal or external nature, will need social support in order to become embedded in new routines (Moos and Schaefer 1986). At the same time, the issue of incentives arises. Alternatives adopted in the short term during a crisis will not be taken on permanently if there is not a clear rationale for doing so. If the coping process has not changed the perceived importance and significance of the fuel crisis to an individual, innovations may be transient. This could have occurred for several reasons including poor communication of the environmental over the economic justification for high fuel prices, or the poor experience users had of public transport. The analysis so far reveals that the potential role the workplace could play in this respect is important.

What does all this tell us? First, that the set of issues that shapes the decision process of road users when choosing modes of transport is clearly quite complex, encompassing personal, social and work considerations. There may be no guarantee that once presented with alternatives to the car all individuals will take advantage of this, however, the challenge is to provide as supportive a context as possible in which road users might consider reducing reliance on the car. In terms of the resilience of innovations, it is clear that the level of change brought about as a result of the fuel crisis was expected to be relatively low. This suggests that the task of reorienting lifestyles and the economy away from the car will be difficult.

However, the process of looking at strategies, contexts and resilience reveals ways in which the lessons of crises may be quickly learned. Crises may create a context for the development of innovative coping strategies and alternative behaviors. Such behaviors may even be preferable to the status quo. Only through looking at the internal dynamics of a crisis, at the level of the individual, family or firm, can these behaviors and alternatives be seen. At the same time, looking at how resilient positive innovations developed during crises might be, can enable us to gain insight into how such innovations might be made more long-term.

Conclusions

This chapter develops an initial framework understanding of how innovation may result from the chaos of technological infrastructural collapse, however temporary, which may enable learning from crisis to occur. The exploratory analysis has revealed that by looking at strategies, contexts and resilience it may be possible to learn lessons from a crisis – by establishing the range of coping strategies and alternative behaviors developed, the way in which the development of innovation in crisis may have been encouraged or inhibited by contextual factors and, finally, the degree to which positive innovations developed in crisis may become embedded in new routines. Such insights into the strategies, contexts and resilience of crisis can only be gained by looking at its internal dynamics, and by focusing not on the return to normality but on any subtle shifts that may have occurred.

Rather than focusing on the disruption caused by reduced access to fuel and the resulting breakdown in the transport infrastructure, this micro-analysis of individuals' actions and alternatives developed to navigate a car-free city has demonstrated the potential for innovation among the chaos. Road users proved themselves able to develop viable strategies for reducing their car usage, and a range of more sustainable and environmentally friendly transport behaviors emerged. In this respect, the fuel crisis created a context for innovation. However, at the same time, employers facilitated or negated the coping efforts of their employees through providing favorable or non-supportive environments for adjustment. In this case, the crisis did not last long enough for new behaviors to become embedded. Nevertheless, this chapter has strong implications for learning from crises and technological collapse, in so far as it offers a framework which stresses the relevance of context in shaping and framing the ability of individuals to innovate and in enabling or hindering such innovations to become embedded in new routines.

Acknowledgments

We would like to thank the 2000 Greater Manchester Common Purpose group for agreeing to participate in this research project.

References

Carpenter, B., ed. (1992) *Personal Coping: Theory, Research and Application*, Connecticut, CT: Praeger Publishers.

Lazarus, R. and Folkman, S. (1984) *Stress, Appraisal and Coping*, New York: Springer.

Lyons, G. and Chatterjee, K. (2002) *Transport Lessons from the Fuel Tax Protests of 2000*, Aldershot: Ashgate.

Moos, R. and Schaefer, J. (1986) "Life Transitions and Crises: A Conceptual Overview," in R. Moos (ed.) *Coping with Life Crises: An Integrated Approach*, London: Plenum Press. 1–28.

Williams, B. (2000) *Financial Times*, 15 September.

PART III
Constructing and Deconstructing the Internet

CHAPTER SIX

Internet: The Social Construction of a "Network Ideology"

Patrice Flichy

During the 1970s and 1980s, the initial designers of the internet were also the first users. Their framework of uses was mainly that of the academic field (for Arpanet) or the counter-culture (for the bulletin board systems). They dreamt of a world where people could exchange information freely from one side of the planet to another, where online communities replaced local communities, and where computerized conferencing afforded the possibility of practising a "collective intelligence." Design, uses and ideologies were unified by the same perceptions in the academic world and the counter-culture.

During the 1990s there was a split in these closed worlds. Designers left the university to work in private companies and the internet became a mass consumption product with widely varied users. A new discourse about network computing and its impact on society was produced by specialists working for computer journals or news magazines. The internet *imaginaire*[1] was no longer that of computer scientists: it had become a mass phenomenon. The *digerati* (digital generation), as the digital intelligentsia called themselves, diffused an internet model of common interest communities, thus creating a "network ideology." But this new internet myth was not completely beyond reality, for the *digerati* were totally familiar with the design of these technologies and their first uses. They acted as mediators between designers and users, organizing the connection but also building the socio-technical framework of the internet. It was they who initiated debate on a digital society.

Virtual Communities: The Founding Myth

In 1993 the internet was put on the media agenda for the first time. At the beginning of the year *Time* magazine published a feature called "Cyberpunk."

After recalling the link between cyber culture and the counter-culture, the article discussed computer viruses, virtual reality, rave parties, drugs (ecstasy) and . . . The Well, the most well-known Californian Bulletin Board System (BBS).[2] In particular, it cited Howard Rheingold, pioneer journalist in the electronic community: "We're replacing the old drugstore soda fountain and town square, where community used to happen in the physical world" (Elmer-Dewitt 1993a: 60). In contrast to those traditional communities, Rheingold spoke of the "virtual community" – the title of his book published at the end of the year.

In September 1993, when the information highway project – debated extensively during the previous year's presidential election campaign – was in a bad way, *Newsweek* published a feature about online life (Kantrowitz 1993). One of the experiences presented was The Well. Rheingold signed an article, an extract from his forthcoming book, in which he emphasized the fact that virtual communities were not utopias since they had actually been created. The book, *The Virtual Community*, was a best-seller,[3] selected by *Business Week* as one of the books of the year. It was the first book about the internet that was neither technical nor a practical manual. The author discussed at length The Well and his own experience as a newsgroup user and host. He also presented other electronic communities such as Arpanet. Through his account he constructed a representation of the net in which virtual communities brought together people from all corners of the globe, many of whom remained attached to their locality. These individuals developed conversations that were as intellectually and emotionally rich as those in real life, in a world of balanced interaction between equals. The Net was presented as helping to recreate a social link and to breathe life into public debate and, more generally, into democratic life.

Rheingold's book thus proposed one of the founding myths of the internet. We know that for Roland Barthes, a myth is in a sense an underlying sign, a meta-language (Barthes 1970: 195–202). He takes as signifier an existing sign and makes it signify something else. Rheingold similarly took the electronic community and invisible college as socio-technical frames and placed them in a sphere different from that of ordinary sociability. In other words, he said that what was good for counter-culture communities, or for universities, was good for society as a whole – as if the change of social sphere would not fundamentally modify the situation. By putting the internet at the heart of contemporary society, a new process of socio-technical construction was inevitably triggered. Rheingoldian mythology overlooked that phase.

Rheingold founded a new utopia. The idea was no longer, as in the 1970s and 1980s, to try out a technical project and to activate small groups of academics around it, but to offer North-American society as a whole the large-scale realization of new communication relationships which, until then, had been experienced in small groups. It is hardly surprising, therefore, that the media used Rheingold as a reference to talk of a universal internet, and that they proclaimed him the "first citizen of the internet."[4] It is similarly understandable that all those who wanted to launch out into the internet business studied The Well, which they considered as the prime example of a virtual community, "hoping to divine the magic formula that made it so special, so captivating, so unique," when, in fact, it was little more than a social experiment (Hafner 1997: 100).

With the publication of Rheingold's book and articles in news magazines, a new internet *imaginaire* appeared. Initially, this may seem to have been a quantitative development. Whereas discourse on the internet had previously been diffused in closed circles (computing, counter-culture, etc.), by more or less confidential media, from 1992–3 it had a place in the mass media. This was the beginning of a classic phase in the development of a technology: mass diffusion following laboratory research and early trials. In this fairly traditional perspective in the sociology of techniques (Rogers 1983), discourse on the new technique is considered simply as a tool to facilitate its diffusion. Yet, the technical *imaginaire* is not something apart from the innovation process, attending it; it is an integral part of it. Let us see how this collective technical imagination was constituted.

Internet for All

Users' Guide

The first literature specifically on the internet for the lay public consisted of handbooks for beginners, first found in bookshops in 1992. These books, written by computer specialists with extensive experience in the new medium, but intended for the general public, offered a wealth of practical information on how to access and surf the web. But they also provided precise representations of this new technique. *Zen and the Art of the Internet* opens with the following statement: "We are truly in an information society." A few lines down the author describes the future internaut:

> You have at your fingertips the ability to talk in "real-time" with someone in Japan, send a 2,000-word short story to a group of people who will critique it for the sheer pleasure of doing so, see if a Macintosh sitting in a lab in Canada is turned on, and find out if someone happens to be sitting in front of their computer (logged on) in Australia, all inside of thirty minutes.
>
> (Kehoe 1992: 3)

Ed Krol was less lyrical:

> Once you're connected to the internet, you have instant access to an almost indescribable wealth of information [. . .] Through electronic mail and bulletin board, you can use a different type of resource: a world wide supply of knowledgeable people, some of whom are certain to share your interests.
>
> (Krol 1992: xix–xx)

He thus conveyed the idea of a sharing of information that was omnipresent in The Well. Moreover, the title of Krol's book (*The Whole Internet User's Guide and Catalog*) was an explicit reference to the *Whole Earth Catalog* and the counter-culture.[5]

These guidebooks catered to a public of academic non-computer specialists and professional users. But Krol foresaw an evolution: "About ten years ago," he wrote, "personal computers brought computing from the realm of technical gurus to the general public [. . .] The internet is currently making the same transition" (Krol 1992: 2) – and his guide had a wide readership, with over 750,000 copies sold. *The Internet Companion*, published in late 1992, was also intended for a larger public constituting "the network community." This community was defined as follows:

> The internet has always been and will always be a key part of the research and development community, but the increase in access and the network's potential for becoming the basis for worldwide communication between people in all walks of life cannot be ignored by the rest of us.
>
> (LaQuey and Ryer 1993: 9)

As an example, a boxed section in the same book recounted the story of two students who met and then married – thanks to the internet! To show how this extension of the internaut community was going to affect everyone, the authors warned their readers in the introduction: "If you want to stay current in the nineties, and even into the next century, you need to learn about the internet." In the following year this broader diffusion was noted by the author of *Internet Starter Kit for Macintosh*: "people are connecting to the internet because the internet is becoming more than just an elite club of technoweenies, it has become a virtual community in and of itself" (Engst 1993: 13). This electronic community, which was attracting an ever-larger public, needed a guide, as Al Gore – then campaigning fervently for information highways – pointed out in the preface to *The Internet Companion*: "for too many people the internet has been uncharted territory, and as a result they have hesitated to explore the vast potential of networking. I trust this book will change that" (LaQuey and Ryer 1993: vi).

All these manuals proposed fairly coherent representations of the internet as a tool for academic work that could be proposed to a broader public for searching for information, producing documents and working with people throughout the world. In this way users could participate in virtual communities, as clearly shown by Mitch Kapor in the preface to another guide available on the net before being published as a book:

> The oldest of these communities is that of the scientists, which actually predates computers. Scientists have long seen themselves as an international community, where ideas were more important than national origin. It is not surprising that the scientists were the first to adopt the new electronic media as their principal means of day-to-day communication.

> I look forward to a day in which everybody, not just scientists, can enjoy similar benefits of a global community.
>
> (Gaffin and Kapor 1991: 8–9)[6]

We find here all the components of the internet myth as described by Rheingold.

Internet Mania

In the autumn of 1993 the major media which had spread the virtual communities myth started to see the internet as a means of mass communication. "Is Middle America ready for the Internet?" asked *Business Week* (Schwartz, E. 1993: 142), while *Time* noted that "suddenly the internet is the place to be" (Elmer-Dewitt 1993b: 62). In the spring of 1994 *Business Week* went further: "hardly a day goes by when there isn't some story in the paper or on the tube hyping the internet" (Baig 1994: 180). These journalists were partly surprised by this success, for the internet culture seemed distant from that of the public at large. *Business Week* cited the president of the online service Prodigy who "calls internet a Wild West show that will never become the backbone of a national data superhighway unless it's usable by mere mortals" (Schwartz, E. 1993: 142). *Time* presented the internet as "an anarchistic electronic freeway" (Elmer-Dewitt 1993b: 62) and in another issue the same journalist, Philip Elmer-Dewitt, cited a consultant for whom "if there is a soul of the internet, it is in that community [of hackers]" (Elmer-Dewitt 1994: 53). This argument was in line with the feature that Elmer-Dewitt had already published the previous year on the cyber-punk. It can be seen as a clear illustration of the ambiguous attitude of *Time* towards the counter-culture. The magazine presented it as a key element in the evolution of American society; it granted it a lot of space[7] and became a sort of mouthpiece for this social movement. Yet, at the same time, it denounced the movement's anti-authority and anarchic behavior. *Time* and the other news magazines consequently changed the way they presented the internet. After the period of promotion of Rheingold's electronic communities came the denunciation of anarchic behaviors. Many articles presented the internet as a medium for pirating and pornography. My study reveals that these two topics constituted roughly half of all articles on the internet in *Time* and *Newsweek*, and about one-eighth of those of *Business Week* in 1993 and 1994.[8]

But the news magazines also fulfilled an educational role by explaining how to use the new medium. "The internet seems to be everywhere, but frankly, you're still puzzled about what it means to you," wrote *Business Week* (Baig 1994: 180). A few months later Katie Hafner wrote a paper in *Newsweek* entitled "Making sense of the internet," with the sub-title "You keep hearing about cyberspace. But what is it, exactly? What's the best way to try it out? And do you really need to care?" (Hafner 1994: 46). For those who did not have the possibility of logging onto the net at work, articles advised people to subscribe to the online services that were to become the main access providers and thus get a new lease on life. The new internet utopia was thus to find a new way to become real.

But the advent of the new internauts often generated conflict. *Internet World*, from 1992 to 1994 the only magazine on the internet for the general public, published a story in November 1994 headed: "Aliens among us. A horde of new users from America Online, Compuserve, Genie and Prodigy is coming onto the internet" (Kantor 1994: 82). This magazine, which published popularized and thought-provoking articles, also wanted to be the mouthpiece of real internauts. In the same issue it published an article denouncing the biased image that the media gave the internet: "judging by what you read in the press, you'd think the internet was primarily a morass of newsgroup in-fighting, purpoiled programs and pictures, retro ex-hippy anti-establishment types and cyber-cowboys and gun-slingers" (ibid.: 99–101).

The internet general public, presented ambivalently by the commercial media, gradually changed. In 1994 *Net Guide* published a description of potential uses of the net under the heading "Your map to the Services, Information and Entertainment on the Electronic Highway," highlighting recreational uses more than any of the guides published in previous years had done:

> you can start relationships, fire off insults, publish your own writings. You can get help on your screenplay. You can get updates on the TV soaps you've missed. You can play games . . . You can search through libraries around the world . . . You can lose yourself in a new medium and a new world.
>
> (Wolf 1994: 1)

A few pages down, the guide answers the question "do I really need it?" with "you'll be part of an active, expanding community of people exploring . . . the information frontier" (ibid.: 10). Once again, this was the same discourse as that of The Well: by participating in this pioneering experience, one tries out a new form of community. Yet, a new orientation of internet was also emerging, owing primarily to the success of the web and the diffusion of Mosaic: a medium for consultation. The advertising banner: "Netguide is the TV guide to Cyberspace" is clearly indicative of this development.

Netiquette

As indicated above, the arrival of new internauts in the forums and, more particularly, on Usenet, often caused conflict due to inappropriate contributions. Yet, the wish to formalize the rules of electronic communications had been expressed on the internet early on. In 1985 two computer scientists who were thoroughly familiar with electronic mail wrote a report at the request of the National Science Foundation, on the ethics and etiquette of electronic mail. The question of etiquette seemed particularly important to them, "because certain standard social norms must be reinterpreted and extended to cover this quite novel medium" (Shapiro and Anderson 1985: 4). These ideas were very soon codified and rules of electronic *savoir-vivre* circulated on Usenet. General principles of any social interaction (identifying oneself, thinking of one's interlocutor, contributing new ideas, etc.) were found alongside rules pertaining to the written media (read before answering, don't answer in the grip of emotion, etc.) or to electronic media (choose the right newsgroup, control the feeling of ubiquity, think that messages are archived, etc.) and, more precisely, to the format of messages (give a title, be brief, etc.).[9] "Net etiquette" was to become netiquette for short. Articles dealing specifically with this question and intended for new users were published, primarily in electronic form. In a biblical tradition, some proposed "the ten commandments for computer ethics" (Rinaldi 1992).

This netiquette gradually became a code of chivalry that real internauts respected and got others to respect. As indicated in *Zen and the Art of the Internet*, "there are many traditions with Usenet, not the least of which is dubbed netiquette" (Kehoe 1992: 43). Virginia Shea, who devoted a book to this question, tried to define the term: "Etymologically, it comes from the French word for 'ticket.' If you know the etiquette for a particular group or society, you have a ticket for entry into it." As she noted further on, the aim of her book was "to give you a 'ticket' to cyberspace. 'Netiquette' is the etiquette of cyberspace" (Shea 1994: 19).

Ed Krol used another metaphor for internet etiquette: "frontier justice." "When the West was young" he wrote, "there was a set of laws for the United States, but they were applied differently west of the Mississippi river. Well, the network is on the frontier of technology, so frontier justice applies here too" (Krol 1992: 135). For these new pioneers, there were two fundamental ethical principles, individualism and the protection of the network, that had to be linked. As soon as the practices of certain users disrupted the normal functioning of a part of the network, pressures of various intensities were exerted on them. Internauts thus practised self-discipline, for "if these problems, instead of finding a solution within the network community, flow over into the press or Congress, no one will stand to gain" (Krol 1992: 45).

To become an internaut it was therefore necessary to abide by the rules of *savoir-vivre* or certain linguistic forms[10] used by the oldest members of the community, academics and researchers. One thus took a "ticket" to join the community or at least to develop practices similar to theirs. Netiquette was designed to promote harmonious debate, moderate controversy and the search for consensus and, on the other hand, to exclude violent debate, which was anathema (sending "flames"). This was a traditional code of conduct in the academic community, considered necessary for the functioning of invisible colleges. By proposing these rules as valid for any electronic social interaction, new internauts' identification with the academic model was reinforced. Netiquette was often mentioned in forums to call to order recalcitrant users and enhance the feeling of belonging to the net. That is what made it different from other codes of conduct. It participated in the symbolic creation of the electronic community and as such was another element in the construction of the internet myth.

Digerati and Cyber-Elite

Wired*: The Cyber Magazine*

Alongside introductory books on the internet and the first articles in the general press, another type of literature appeared which developed a far broader *imaginaire* concerning the entire computing world. In January 1993, when the public at large was timidly starting to link up to the net and information highways were at the center of the political agenda, a new magazine was launched in San Francisco: *Wired.* Like their first sponsor, Nicholas Negroponte (director of MIT's Media Lab), the founders, Louis Rossetto (managing editor) and Jane Metcalfe (president), were persuaded that computer technologies were going to trigger a real revolution. Apart from the universal internet that some were promoting, computing was infiltrating all human activities, professional and personal, intellectual and artistic alike. To the question asked in the editorial headed "Why *Wired*?," Rossetto answered:

> because the Digital Revolution is whipping through our lives like a Bengali typhoon . . . and because the computer "press" is too busy churning out the latest PCInfoComputingCorporateWorld iteration . . . to discuss the meaning or context of social changes so profound their only parallel is probably the discovery of fire . . . *Wired* is about the most powerful people on the planet today – the Digital Generation.

The tone was set: *Wired* was to be the standard-bearer of the new computer culture. As Metcalfe later said:

> What we are really talking about is a fundamental shift in society that is being led by technology but is infiltrating every aspect of society . . . *Wired* is really about change. It's led by technology, absorbed by business, and spread by artists. But it's not about technology.
>
> (quoted in Brockman 1996: 221)

This new magazine was not only the tribune for new ideas on the computer revolution, it also had a new layout that was a milestone in the history of press magazines. Articles were presented in the same way as advertisements, with slogans over photos or omnipresent graphic illustrations. Photos were often manipulated and represented anything from landscapes or people to printed circuits or computer terminals. The picture rarely had an illustrative function; it was in constant interaction with the text, sometimes even making it difficult to read. This layout was similar to that found shortly afterwards on web sites or CD-ROMs. It was even judged sufficiently innovative to be exhibited in the San Francisco Museum of Modern Art. *Wired* was more like *Rolling Stone* than like an intellectual journal, and this resemblance was not only formal. It was also a standpoint deliberately adopted in the cultural field. *Newsweek*, which announced the launch of the magazine, noted that Rossetto's aim was to create the *Rolling Stone* of the computer generation (Schwartz, J. 1993: 62). Two years later Paul Keegan noted: "like *Rolling Stone* in the 1960s, it has become the totem of a major cultural movement" (Keegan 1995: 39).

One of the appealing aspects of the magazine related to the fact that Rossetto managed to gather around him a group of intellectuals and journalists who came to be known as the cyber-elite. Kelvin Kelly, former editor-in-chief of the *World Earth Review*, occupied the same position at *Wired*. Stewart Brand, founder of the *World Earth Review* and of The Well was also on the editorial committee, as was Rheingold. Many editors were members of The Well.

We also find journalists from other backgrounds, such as Steven Levy, editorialist with *Newsweek*, Joshua Quittner, journalist with *Time*, John Markoff, correspondent for the *New York Times* in Silicon Valley and R.U. Sirius, editor-in-chief of the New Age magazine *Mondo 2000*. Two science fiction novelists, William Gibson – the first person to use the term "cyberspace," in his novel *Neuromancer* (1984) – and Bruce Sterling, also participated in the undertaking, as did creators using new technologies, such as Jaron Lanier and Brenda Laurel. Last, consultants, specialists in forecasting, such as Esther Dyson and Paul Saffo, were also there. In fact, forecasting had a place of honor in the magazine. For example, the fifth issue featured a long interview with Alvin Toffler whose thinking directly inspired Rossetto:

> This is the mainstream culture of the twenty-first century. It's a new economy, a new counterculture and beyond politics. In ten or twenty years, the world will be completely transformed. Everything we know will be different . . . I think Alvin Toffler's basically right: we're in a phase change of civilizations here.
>
> (quoted in Keegan 1995: 39)

Rossetto's esteem for the futurologist seems to have been mutual, for Toffler commented: "I think the readers of *Wired* are a unique resource for the country" (Kelly 1996). And indeed, as Paul Keegan noted: "the genius of *Wired* is that it makes the Digital Revolution a self-fulfilling prophecy, both illuminating this new sub-culture and promoting it – thus creating new demand for digital tools, digital toys, digital attitudes" (Keegan 1995: 40).

Rosetto and Metcalfe had a lot of difficulty finding capital to launch their magazine. Many publishers thought the market was too limited. As one of them said: "if you get out of Silicon Valley there are not a lot of places where you find that psychographic group" (Schwartz, J. 1993: 62). But contrary to all expectations, *Wired* immediately had a vast readership: 110,000 *digerati* after one year in existence (Kantrowitz 1994: 38), 230,000 after two years (Keegan 1995: 39) and 450,000 after four (*San Francisco Chronicle*, 8 May 1998: A20), that is, nearly half the readership of the first popularized technical magazines such as *PC World*. These readers were people with a passion for the internet and other digital media. We find among them many professionals of these new technologies: computer scientists, multimedia designers, artists, etc.[11]

Forecasting Club and Association to Defend Computer Freedoms

The cyber-*imaginaire* appeared not only in the media but also in think-tanks. In 1988 Stewart Brand, along with futurologists who had worked for Shell and graduates from Stanford Research Institute, formed a forecaster's club which they called the Global Business Network (GBN). This institute sold services to firms and administrations. As one of its clients commented: "the network is a curious blend of scientists, musicians, artists, economists, anthropologists, and information technology gym rats who form a mosaic by which us capitalists can view our business environment and even our company" (quoted by Garreau 1994: 157). GBN was thus to act as an interface between the business world and heterodox future scenarios in which information technologies and ecology were to have prime importance. Apart from a small core of permanent members, the institute used a network of experts. It used The Well to organize a private teleconference between its experts and to host its presentation on what would later be called a web site. *Wired* developed close ties with GBN and the magazine's forecasts owe a lot to the institute. About ten members of GBN were on the magazine's editorial committee. Four of them were on the cover pages of the 17 first issues.

While GBN developed forecasting on the information society and provided a link with thinking in the business world, another, more political, line of thought was established with the Electronic Frontier Foundation. At the origin of this association lay an event that marked the computer specialist community. In 1990 the FBI launched an inquiry that implicated a large number of hackers suspected of pirating software. John Barlow, former hippie (lyric writer for Grateful Dead) and computer journalist, Mitchell Kapor, co-founder of the computer company Lotus (later sold), and John Gilmore, another hacker who had got rich from computing,[12] decided to found an association to defend freedom in cyberspace. The Electronic Frontier Foundation (EFF) was:

> established to help civilize the electronic frontier; to make it truly useful and beneficial not just to a technical elite, but to everyone; and to do this in a way which is in keeping with our society's highest traditions of the free and open flow of information and communication.
>
> (EFF undated)

More precisely, it defended individual cases and its Washington office lobbied the government and Congress. Kapor, an habitué of The Well, naturally opened a conference on the foundation in the Californian BBS. The founders, moreover, considered The Well to be "the home of the Electronic Frontier Foundation" (Sterling 1993: 238).[13] The EFF was also complementary to *Wired*. It provided the magazine with one of its main topics and *Wired*, in turn, constituted an excellent tribune for the association. Two of the three founders, half the board of directors and the legal adviser (Mike Godwin) wrote in the magazine. Jane Metcalfe was also elected to the board.

Thus, The Well, the Global Business Network, the Electric Frontier Foundation and *Wired* all had a lot in common. By clicking on the name of one of the actors of these institutions, one often found oneself in another institution, in the hypertext of the cyber imagination. Referrals were constant as each institution supported the others. GBN contributed its forecasting and its contacts with big business, EFF its political project and struggle to defend the freedom of internauts, *Wired* its ability to format and adhere to the cyberpunk culture. Stewart Brand, one of the main links between these institutions, summarized the relations of all these projects with the counter-culture when he noted: "one advantage of working with survivors of the 1960s is, we've had an experience in creating utopias. We've had our noses rubbed in our fondest fantasies" (quoted by Quittner 1994: 140).

The editors of *Wired* and members of the Electronic Frontier Foundation or of the Global Business Networks were not only, like the editors of popularized manuals, the agents of the diffusion of internet. They saw themselves above all as an elite, as the avant garde of the information society. Different lists of the people who would count in the future information society were published in the press.[14] Journalist and literary agent John Brockman published a *digerati* "Who's Who" in 1996. The 40 or so people selected included computer scientists, entrepreneurs and intellectuals, among others (about ten editors of *Wired* were included). As Stewart Brand said: "elites are idea and execution factories. Elites make things happen; they drive culture, civilization" (Brockman 1996: xxxi).

Thus, this cyber-elite included many intellectuals and journalists. Although these specialists of discourse were not cut off from innovators, it was nevertheless they who, in the mid-1990s, produced the collective vision of the information society. The *digerati*'s discourse contained the new forms of politics, economy and definition of the self that emerged with the digital revolution.

The *digerati*'s ideology was finally spread by the mass media in 1995. The editorialist of *Time* noted that:

> most conventional computer systems are hierarchical and proprietary; they run on copyright software in a pyramid structure that gives dictatorial powers to the system operators who sit on top. The internet, by contrast, is open (non proprietary) and rabidly democratic. No one owns it. No single organization controls it. It is run like a commune with 4.8 million fiercely independent members (called hosts). It crosses national boundaries and answers to no sovereign. It is literally lawless . . . Stripped of the external trappings of wealth, power, beauty and social status, people tend to be judged in the cyberspace of internet only by their ideas.
>
> (*Time* special issue, March 1995: 9)

Newsweek made 1995 Internet Year. It opened its year-end issue with the following phrase spread across four pages "this changes . . . everything," and the editorial described the internet as "the medium that will change the way we communicate, shop, publish and (so the cybersmut cops warned) be damned" (*Newsweek* special issue, 2 January 1996).

Conclusion

These discourses impacted strongly on the future of the internet. In fact, they proposed a framework of interpretation and action for network computing and showed what could be done with the internet and how. This frame was all the more powerful in so far as it described communication practices that functioned in the academic and counter-culture worlds, and to which access could be organized. By becoming a new internaut one not only became a user of network computing and of communication or information retrieval tools; one also entered into another social world where relations between individuals were equal and cooperative, and information was free.

This view is, indeed, somewhat strange, for society is neither a "cybercampus" nor a "cybercommune." Inequalities in skills (in the use of computing and the production of discourse), of a far greater dimension than in the academic world, have appeared. The principle of gratuity has faded with the need to finance certain resources through media-type means (subscriptions, advertising, etc.). But the initial model has nevertheless lasted. Forums for the public at large have been set up, information collated by universities is consulted by different users, and ordinary individuals create sites where they present information that is sometimes of great value. Thus, this model of communication is, in spite of all, a framework for interpretation and action that is only partly unsuited to the new reality of the internet. During the 1990s it provided a range of uses and behaviors on the internet.

Notes

1 Myth, utopia, ideology or collective vision product of the collective imagination.
2 Following this article, The Well hosts received many messages asking them: "Is this the cyberspace?" (Dery 1994: 6–7).
3 At first 35,000 copies were printed. The following year the book was printed in paperback.
4 An expression found in one of the critiques in the presentation of his book on the Amazon.com site.
5 A far shorter version of this guide had been published in 1989 under the title *The Hitchhiker's Guide to the Internet*, in the framework of Requests for Comments (no. 1118) on the internet.
6 This guide was published in the form of a book under the title *Everybody's Guide to the Internet* (Cambridge, MA: MIT Press, 1994).
7 In March 1995, *Time* published another special issue entitled "Welcome to Cyberspace" featuring an article by Stewart Brand with the title "We Owe it All to the Hippies" and the sub-title "Forget antiwar protests, Woodstock, even long hair. The real legacy of the sixties generation is the computer revolution."
8 *Time*: 9 articles of which 3 on sex and 2 on pirating. *Newsweek*: 14 articles of which 1 on sex and 6 on pirating. *Business Week*: 8 articles of which 1 on pirating.
9 One of the first presentations on paper of these rules are found in Quarterman 1990: 34–7.
10 Smileys, symbols indicating the emotive weight of a word, are another way of showing one's familiarity with the internet.

11 The average reader was 37 years old and has an income of US$122,000 (source: *San Francisco Weekly*, 1996).

12 John Gilmore was one of the first employees of the computer company Sun Microsystem, and was paid mostly in stock options. A few years later he sold his shares and was able to live off the interest.

13 On the history of the Electronic Frontier Foundation, see also H. Rheingold, *The Virtual Comunity* (1994): 256–60.

14 See, for example, "The Net 50." "They're supplying the vision, the tools and the content that are getting millions of people to turn on their modems" (Anonymous 1995: 42–6).

References

Anonymous (1995) "The Net 50," *Newsweek*, 25 December: 42.

Baig, E. (1994) "Ready to Cruise the Internet?," *Business Week*, 28 March: 180–1.

Barthes, R. (1970) *Mythologies*, Paris: Le Seuil.

Brand, S. (1995) "We Owe it All to the Hippies," *Time*, 1 March: 54–7.

Brockman, J. (1996) *Digerati. Encounters with the Cyber Elite*, San Francisco, CA: Hardwired.

Dery, M. (1994) *Flame Wars*, Durham, NC: Duke University Press.

Electronic Frontier Foundation, *Mission Statement*. Online. Available at: http://www.eff.org/EFFdocs/about_eff.html (accessed February 2001).

Elmer-Dewitt, P. (1993a) "Cyber Punk," *Time*, 8 February: 58–65.

Elmer-Dewitt, P. (1993b) "First Nation in Cyberspace," *Time*, 6 December: 62–4.

Elmer-Dewitt, P. (1994) "Battle for the Soul of the Internet," *Time*, 25 July: 50–6.

Engst, A. (1993) *Internet Starter Kit for Macintosh*, Indianapolis, IN: Hayden Books.

Gaffin, A. and Kapor, M. (1991) *Big Dummy's Guide to the Internet*. Online. Available at: http://www.thegulf.com/InternetGuide.html (accessed February 2001).

Garreau, J. (1994) "Conspiracy of Hertics," *Wired*, November: 98–104.

Hafner, K. (1994) "Making Sense of the Internet," *Newsweek*, 24 October: 46–8.

Hafner, K. (1997) "The Epic Saga of The Well," *Wired*, May: 98–142.

Kantor, A. (1994) "Aliens Among Us," *Internet World*, November/December: 82.

Kantrowitz, B. (1993) "Livewires," *Newsweek*, 6 September: 42–9.

Kantrowitz, B. (1994) "Happy Birthday: Still Wired at One," *Newsweek*, 17 January: 58.

Keegan, P. (1995) "The Digerati! Wired Magazine has Triumphed by Turning Mild-mannered Computer Nerds into a Super-desirable Consumer Niche," *New York Times Magazine*, 21 May: 38–45, 86–8.

Kehoe, B.P. (1992) *Zen and the Art of the Internet. A Beginner's Guide to the Internet*, Englewood Cliffs, NJ: Prentice Hall.

Kelly, K. (1996) "Anticipatory Democracy," an interview with Alvin Toffler in *Wired*, July.

Krol, E. (1992) *The Whole Internet User's Guide & Catalog*, Sebastopol, CA: O'Reilly & Associates.

LaQuey, T. and Ryer, J. (1993) *The Internet Companion. A Beginner's Guide to Global Networking*, Reading, MA: Addison-Wesley.

Quarterman, J. (1990) *The Matrix*. Bedford, MA: Digital Press.

Quittner, J. (1994) "The Merry Pranksters Go to Washington," *Wired*, June: 77–80.

Rheingold, Howard [1993] (1994) *The Virtual Community*, New York: Harper Collins.

Rinaldi, A. (1992) *The Net. User Guidelines and Netiquette.* Online. Available at: http://www.listserv.acsu.buffalo.edu/c...A2=ind9207&L=nettrain&F=&S=&P=1383 (accessed February 2001).

Rogers, E. (1983) *Diffusion of Innovations*, New York: Free Press.

Schwartz, E. (1993) "The Cleavers Enter Cyberspace," *Business Week*, 11 October: 142.

Schwartz, J. (1993) "Propeller Head Heaven. A Technie Rolling Stone," *Newsweek*, 18 January: 62.

Shapiro, N. and Anderson, R. (1985) *Towards an Ethics and Etiquette for Electronic Mail*, Santa Monica, CA: Rand Corporation.
Shea, V. (1994) *Netiquette*, San Francisco, CA: Albion Books.
Sterling, B. (1993) *The Hacker Crackdown*, New York: Bantam.
Wolf, M. (1994) *Net Guide*, New York: Random House.

CHAPTER SEVEN

The Diffusion of Information and Communication Technologies in Lower-Income Groups: *Cabinas de Internet* in Lima, Peru

Ana María Fernández-Maldonado

During a conference at the headquarters of the Inter-American Development Bank in Washington DC in May 2002, the former Peruvian Prime Minister Roberto Dañino remarked on the importance of the internet for the poorer groups of society, explaining that in Peru, such groups were using internet technology to improve their quality of life. He mentioned that three out of four internet users in Peru accessed the internet via *cabinas públicas de internet* (Drosdoff 2002), small-scale storefront businesses, similar to cyber cafés, which offer low-cost, reliable connections to the internet and which have become a familiar urban facility in middle- and low-income neighborhoods.

This proud announcement might lead us to think that the establishment of these *cabinas* was the fruit of a deliberate government initiative or a well-organized policy for universal access to the internet in Peru. But this has not been the case. While other Latin American countries with a high rate of internet access have relied on government intervention, Peru's government has offered no support other than to regulate competition in the telecommunications sector. Rather, the development of *cabinas* has resulted from thousands of private initiatives by small, local entrepreneurs, who have been quick to understand that there was a high demand for information and communication technologies (ICTs) that could not be met under the traditional system of individual ownership under current economic conditions. Peru thus constitutes an interesting case of diffusion of ICTs in low-income population groups.

The main purpose of this chapter is to document the diffusion of ICTs in the different socio-economic sectors in Lima and to gain a better understanding of the uses and development of ICTs, while paying special attention to lower-income groups or sectors. The technologies considered are fixed and mobile telephony, cable TV, PCs and, especially, the internet. The discussion is based on the results of several surveys carried out in Metropolitan Lima on ICT uses at home and in *cabinas*, and on interviews with people working in the field. The first part of the chapter discusses the patterns of the social diffusion of internet use among Lima's population. The second part focuses on the use of *cabinas de internet*. The final part presents a critical analysis of the significance of ICTs for the improvement of the daily life of poor Lima residents.

General Diffusion of ICTs

Social Polarization in Lima: Origins and Current Trends

Lima is the fifth largest Latin American city, resembling a typical third world metropolis. With nearly 8 million inhabitants, it contains 30 percent of the population of the country. Lima continues to grow at a high annual rate and accounts for more than 50 percent of the economic activity of Peru. Circumstances have turned Lima into a socially divided city, deeply affected by the process of urbanization that has triggered massive rural–urban migration, beginning in the 1950s. Although at present most of Lima's growth is generated internally, 62 percent of the heads of households living in Metropolitan Lima in 2000 were born outside the capital (El Comercio y Apoyo 2001: 246).

Lima has been increasingly unable to provide employment, housing, and urban services for the majority of its citizens. Newcomers to the city, lacking urban services and amenities, have sought to obtain them through their own individual or collective efforts. The results are visible. Most residents now live in sub-standard dwellings located in neighborhoods without essential urban facilities. Only 22 percent of the city's heads of households are registered for tax (Apoyo 2000b: 8) and most workers have jobs in the informal sector. New informal services emerge each day and public transportation is in the hands of the informal sector, making Lima more chaotic than ever before, but at the same time providing an indispensable service that is affordable to the majority of residents. This seems to be the fate of Lima: as a result of decades of *laissez-faire* policies only the informal sector is able to provide some of the services residents need at affordable prices.

More than other Latin American large cities, Lima has suffered from deep political and economic instability since the beginning of the 1980s. The Peruvian economy collapsed during the so-called "lost decade of Latin America," after decades of industrially oriented growth. The crisis dramatically affected poorer groups and led to the emergence of violent political movements that made Lima the center of their operations for several years.

In 1990, a new government initiated a plan aimed at restructuring the economy and minimizing state intervention. At the same time it succeeded in fighting the insurgency movements. However, the processes of wholesale privatization, reduction of the public sector and other liberal policies have had tremendous social costs, adversely affecting the daily life of the average Peruvian. Poverty has increased and inequalities have become more acute. The effects of the Asian crisis and the El Niño phenomenon resulted in a new and deeper recession that began in late 1998 and affected the national economy until 2001.

The long cycle of economic recession changed the composition of socio-economic levels in Lima, dramatically increasing the proportion of persons living in poverty, which represented 48.8 percent of the total population in 2000. The socio-economic composition of the population in 1991 and 2000 (see Table 7.1) clearly reveals the increasing polarization of society during the 1990s. The middle-income (B) sector declined by more than half a million people in absolute terms during this period, producing what has been called the process of the "dissolution of the middle-class." Average monthly household income in 2000 was US$409 (the median was US$235), which was 10 percent lower than in 1999 and 23 percent lower than in 1998 (Apoyo 2000b). This dramatic reduction in income undoubtedly forced most people to reorganize their expenditure and consumption habits.

The dramatic changes in the economy during the 1990s resulted in major transformations in the functioning of the city, which in turn have produced great physical transformations. In the late 1990s, real-estate investment groups pumped unprecedented amounts of capital into new projects. On the other hand, in other parts of the city, the increasing numbers of Andean migrants have broken down important economic and social barriers and have become a much-appreciated new market and the driver of many new urban trends. They have reinforced the pattern of multiple centers observed in Lima during the 1980s, producing a more dispersed and decentralized urban structure. Ludeña (2001: 17) describes the changes as a "contradictory process of democratization and social exclusion in the use and development of the urban space."

As a result of a decade of "wild capitalism," Lima begins the twenty-first century more than ever a city full of contradictions and contrasts. The spatial structure of the city can no longer be described in terms of a dichotomy between the formal and informal, or rich and poor areas. In terms of its economic dynamics, Lima remains a basically informal city, excluded from the global financial circuits. It is within this context that the telecommunications revolution has affected Lima and a real internet boom has been experienced in an incredibly short period. "Lima emerges in the year 2000 dotted with *cabinas públicas*, *cybercafés* and cybershops" (El Comercio y Apoyo 2001: 214).

Table 7.1 *Changes in Socio-Economic Levels in the Population of Lima 1991–2000*

	High-Income Sector (A) %	*Middle-Income Sector (B) %*	*Low-Income Sector (C) %*	*Very Low-Income Sector (D) %*
1991	3.8	21.0	38.7	36.5
2000	3.8	14.3	33.1	48.8

Source: Data adapted from Apoyo 2000b.

General Diffusion of ICTs in Lima

The diffusion of ICTs in Lima reflects the urban situation of the city and is characterized by an intriguing combination of "digital exclusion" and democratization.

On the one hand, despite the expansion and modernization of the telecommunications networks that followed the privatization and liberalization of the Peruvian telecommunications industry after 1994, access to telecommunications remains very limited compared to the well-connected countries of the North, and relatively low compared to other countries in Latin America. In 2001, 70.8 percent of Peru's population had no telecommunications services at all. This proportion increased to 93.8 percent for the rural population and decreased to 57.5 percent for urban residents (INEI 2001). The density of telephone coverage was 6.3 lines per 100 inhabitants in June 2003 (OSIPTEL 2003), while the Latin American average was 12 lines (ITU 2000). Furthermore, telephone diffusion is highly skewed, with far more lines per inhabitant in the capital city than in the rest of the country. The telephone, and ICTs in general, are also skewed from a social perspective, with high levels of diffusion in the higher income groups, which constitute less than 20 percent of the total population, and low levels in the poorer sectors. The government has done surprisingly little to favor the diffusion of ICTs, particularly among poorer groups.

On the other hand, mobile telephony and internet access have increased visibly over the last few years. Mobile telephone density was 9.2 users per 100 inhabitants in June 2003 (OSIPTEL 2003). Moreover, despite the economic limitations to home diffusion of ICTs, the growth in internet users in Metropolitan Lima in recent years has been astonishing. This is especially true in the city of Lima where 23 percent of the total population were internet users in June 2002 (Apoyo 2002a). If we consider the group consisting of men and women between 12 and 50 years old, 37 percent of them used internet at least once a month in 2002, while in 2000 the proportion was only 18 percent (Apoyo 2002a). Since the average Lima resident cannot afford the basic tools for home digital connection – a telephone line and a computer with a modem – people are accessing the internet in public places. Figure 7.1 shows the distribution of users according to the places where they access the internet in Metropolitan Lima, and the changes that have occurred between 2000 and 2002. The proportion of internet users that accessed the net from the *cabinas públicas* increased from 76 to 89 percent during that period, while home access decreased from 17 to 11 percent (Apoyo 2000a, 2001, 2002a).

In August 2000, the National Institute of Statistics and Informatics (INEI) carried out a survey of ICT use by household in Metropolitan Lima. Table 7.2 indicates the proportion of Limenean households with PCs or computer skills, according to household income quintiles. The results with respect to PC skills are surprisingly high in light of the rate of PC ownership. Even in the first quintile (the poorest households) the presence of members with PC skills was frequently observed. The comparison between the high levels of PC skills and the low levels of PC ownership clearly suggests that many individuals are using ICTs

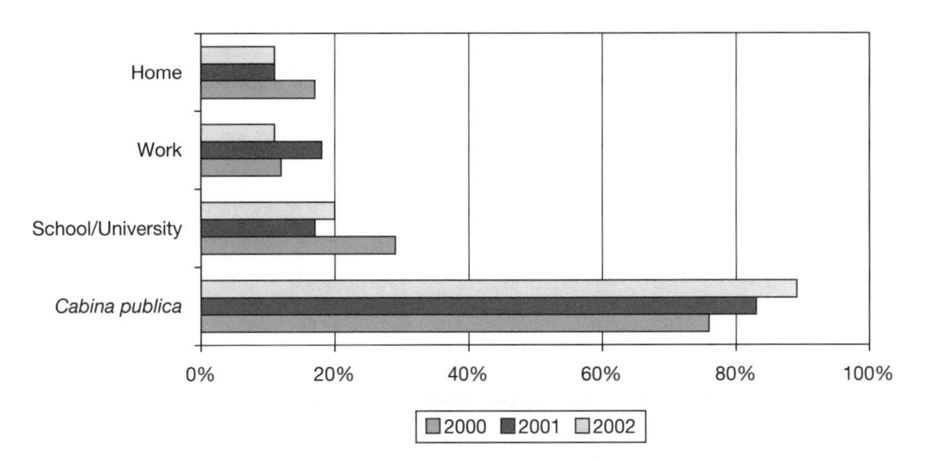

Figure 7.1 *Place of Access to the Internet in Metropolitan Lima in 2000, 2001, and 2002*
Source: Data from Apoyo 2000a, 2001, 2002a.

outside their home. Furthermore, the gap observed between rates of connection to ICTs and of PC ownership is much wider than the gap between rates of connection to ICTs and ICT skills.

The proportion of households with PC skills would probably be lower without the *cabinas.* The role of the *cabinas* in democratizing internet access can also be inferred from the results of surveys on internet use in Metropolitan Lima which illustrate the exceptional growth in users during the last two years, a period which coincided with the proliferation of *cabinas* in Lima neighborhoods. The results also show that "newcomers" to the internet, i.e., users with less than one year's experience, mainly belong to the poorest socio-economic sector (group D, see Table 7.3).

During the same period, the number of home internet connections increased very slowly. This lag is basically linked to the high price of telephone services.

Table 7.2 *PC Skills and PC Possession in Metropolitan Lima Households According to Income Quintile in 2000*

	PC Skills %	*PC Possession* %
I quintile (poorest 20%)	37.8	0.2
II quintile	55.1	4.2
III quintile	65.6	6.8
IV quintile	64.8	11.3
V quintile	77.0	32.8

Source: Data adapted from INEI 2000.

Table 7.3 Experience in Internet Use According to Socio-Economic Sector

	High-Income Sector (A) %	Middle-Income Sector (B) %	Low-Income Sector (C) %	Very Low-Income Sector (D) %
Three or more years	37	39	27	11
Two years	24	26	27	26
One year	13	19	27	31
Less than a year	11	13	19	31
Average	2.7 years	2.4 years	1.9 years	1.3 years

Source: Data adapted from Apoyo 2002a.

Price is clearly the most significant barrier to further development of ICT diffusion, beginning with telephonic diffusion, which in turn affects internet diffusion. Forty dollars for an (average) domestic internet connection represents a luxury that few can afford and the average use of a dial-up internet connection (including internet and telephone charges), or a cable-modem connection, costs approximately the same.

A second barrier is the limited awareness of the political classes as to what is at stake with the introduction and development of new technologies. Despite the media hype and proclaimed affinity with ICTs, there is still no strategic vision about the role that these technologies should play in national or local development. For years, the political instability and economic difficulties of the country hindered the development of national or local policies to promote ICT diffusion. A first initiative was launched in 2001 with the Huascaran Plan, aimed at developing ICTs in public schools.[1] Its strategic orientation relates to educational content, training and the establishment of a technological platform to link schools and educational institutions by means of modules. The objective is to build 5,000 modules over the next five years. However, this plan requires substantial financial resources, which are difficult to find in these times of economic recession. Moreover, there are no other significant government initiatives to promote access, although the present government has recently shown some interest in *cabinas*.

Finally, there is another less specific obstacle to the diffusion of ICTs related to the new rules of the game in the telecommunications sector. Because of the removal of state support, the construction of the information society in Peru basically responds to the commercial interests of the big telecommunications companies operating in the country, most of which are in foreign hands. This favors sectors of high demand and expected profitability.[2]

Figure 7.2 shows average ICT diffusion according to socio-economic sectors in Lima in 2001. The graph clearly illustrates the high degree of "digital polarization": high- and middle-income groups (A and B) are well served, but low-income groups (C and D), representing more than 80 percent of the total population of Lima, are clearly neglected.

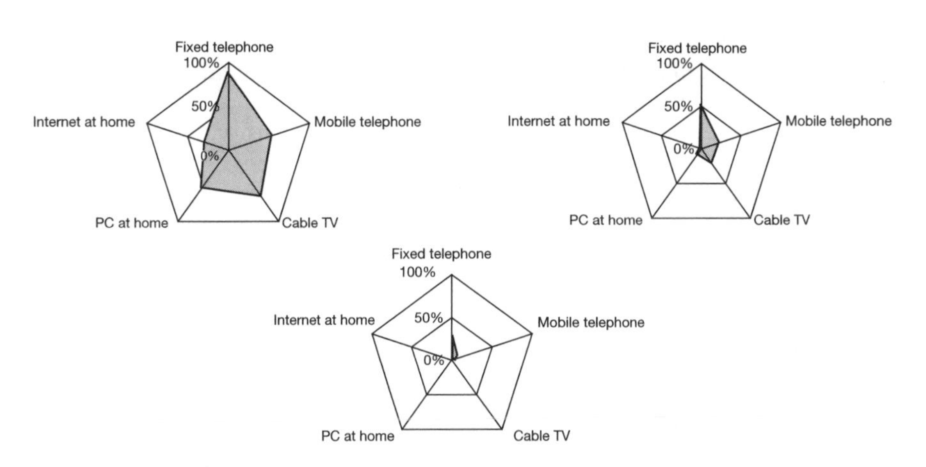

Figure 7.2 *ICT Diffusion in Metropolitan Lima in 2001 by Socio-Economic Level. Top left: A and B sectors. Top right: C sector. Bottom: D sector*

Source: Data from OSIPTEL 2003.

The *Cabinas Públicas*: Diffusion and Use

A New Informal Service in the City

The previous section showed that lower-income sectors and groups have minimal home access to information and communications technologies. However, lower-income groups have a certain level of access to telecommunications services thanks to the *cabinas públicas de internet* that have popped up over the city. The informal sector has effectively acknowledged the high demand for telecommunications services in lower-income groups and has entered the business with great success. This process began at the end of 1998, coinciding with the beginning of the last recession, when 10 percent of telephone lines in Lima (approximately 150,000 lines) were given back to the phone company because they had become unaffordable for users (Fernández-Maldonado 1999).

Indeed, the informal sector began to get involved in ICT-related businesses long before the emergence of the *cabinas públicas*. Retailing in ICT-related products, renting computers and peripherals, and providing informal computer services were part of the initial trade. It was at this time that the potential of the *cabinas* was identified. Thanks to the presence of this informal market for ICT services and products and to the abilities of the informal sector to work on very low revenues, the informal entrepreneurs have been able to meet the demands of low-income groups.

The *cabinas públicas* do not differ much from internet cafés in other cities: they are places where computers connected to the internet are rented for a fixed price per hour. What makes them special is that the prices are set in accordance with local living standards, which are not very high, and that there are a great number of them, distributed throughout the city and serving local residents. However, they are not part of a centralized administration or organization; they are, rather, the fruit of thousands of commercial initiatives by small (informal) entrepreneurs.

The computers are connected to the internet by means of a dedicated line with a monthly subscription to an internet service provider (ISP). The *cabinas* are characterized by their low prices and relatively efficient connectivity: the connection with a dedicated line is much quicker and less problematic than through a phone line shared with the internet, as is the case in domestic connections. *Cabinas* are also multifunctional. They began by offering traditional internet services such as e-mail and access to the web, however they have subsequently diversified their business according to local demand. Thus, in 2002, 67 percent of *cabinas* offered office services such as faxing, scanning, printing, photocopying, and text editing; 55 percent provided communication services through internet, such as long-distance calls and video-conferencing; 41 percent also offered sweets and drinks; some *cabinas* offered totally private modules so that clients can enjoy privacy, while others offered additional services such as extra fast connections and CD-writers. An additional feature of the *cabinas* is their ease of use: unskilled users can rent a computer and they will generally receive assistance in operating computer programs and basic applications (Apoyo 2002a).

Cabinas originally operated only in the center of the city or in middle-income districts, however, since 1998, they have spread to the rest of the city, including the so-called Cones, the peripheral areas where 60 percent of Lima's population live, and which form the basis of Lima's informal economy (70 percent of the dwellings in the Cones are constructed by the inhabitants themselves, informally) (Apoyo 2000c: 41). The percentage of internet users that visit *cabinas* in the three Cones and the city center is higher than in the better-off areas of the city in the south-east and the western districts.

Since they have spread to poorer neighborhoods, *cabinas* have become popular places to gather informally; they are the new public plazas where youth meet. In a recent study, 44 percent of users surveyed reported being accompanied to a *cabina* by a friend or relative; this was especially true of young users. Most of them (75 percent) visited a *cabina* that was within walking distance of their house. Only 10 percent of users surveyed visited *cabinas* outside their districts (Apoyo 2002a).

Because of their success, the *cabinas* are increasingly attracting the interest of public and private institutions. Government officials are now lauding the advantages of the *cabinas* model in international forums and congresses. A successful online forum on "internet *Cabinas*: opportunities for all" was organized in December 2002 by the National Council of Science and Technology (*Consejo Nacional de Ciencia y Tecnología* – CONCYTEC) for the purpose of discussing the main problems and possibilities of the *cabinas* with the *cabinas* owners, administrators, users and other stake-holders.[3]

The government recently approved new legislation to regulate the functioning of the *cabinas*. According to this legislation, all *cabinas* should have at least two computers suitable for use by children. Furthermore, all computers should have filters to avoid children being exposed to adult content. Some *cabinas* have been closed by local authorities for not complying with the new regulations.

Private firms have also identified business possibilities in the *cabinas*. Traditional and online newspapers and portals pay close attention to their development and publish lists of where people can find them. Some of these (*El Comercio*, *Adonde*, *Terra Networks*, *RCP*, etc.) have established networks of *cabinas* for different purposes. New firms have also been established to serve the needs of *cabinas* owners. The *Revista Info Cab*, a specialized magazine, is distributed to 2,500 *cabinas* in Metropolitan Lima. Other firms have organized meetings, congresses, and fairs on *cabinas*.

Who Uses Cabinas and How Do They Use Them?

Various studies of *cabinas* undertaken in Lima have helped characterize users and their reasons for visiting these establishments. Most of these surveys were carried out in low-income areas.

The first study of the *cabinas* phenomenon was conducted in January 1999 by F. Nagaro. He interviewed 200 people in five *cabinas* located in Wilson, the informal ICT market in the center of Lima, and in Villa El Salvador in the South Cone (Nagaro 1999). There were approximately 200 *cabinas* in Lima at the time of the study (Fernández-Maldonado 1999). While users of *cabinas* in Wilson also came from different districts, all users in Villa were residents of the area. Most users in the study were young students; the average age in Wilson was 23.1 (of which 62 percent were students) and 20.3 in Villa (of which 74 percent were students). A high proportion of women used the *cabinas* in Villa (60 percent). Thirty-five percent of all users had computers in their homes, but only 4 percent had internet connections, most of them from Wilson. Remarkably, 11 percent of the *cabinas* interviewees did not use computers for the internet and 45 percent used them for that purpose only occasionally. Indeed, the first *cabinas* originated (in Wilson, precisely) as places which rented computers which were not necessarily connected to the internet (Fernández-Maldonado 1999). In these cases, the computers were rented to produce documents for academic or school purposes, to lay-out texts, design web pages, etc. On the other hand, chat groups and e-mail were the most popular uses among those connected to the internet. Chat groups were especially popular among the young. Another common use was to practice running different programs in the hope that it would improve the user's job marketability.

A second study was conducted a year later in January 2000 in 25 *cabinas públicas* in Surco and Villa El Salvador (a middle-class and a low-income district, respectively) (Fernández-Maldonado 2000). By the time of the second study, the number of *cabinas públicas* in Lima had increased to approximately 1,000 (Lama 1999). During this second study, a more diversified group of users (children and older people) was observed, although young people continued to be the dominant group. A remarkable feature was the rapid and visible increase in demand during this period: the *cabinas* were crowded with clients every day of the week. However, the most remarkable change since the first study was the now widespread use of the internet as a communications and information tool. Almost all computers were connected to the internet, and almost all users were using the internet actively. E-mail (basically through web-mail) and chat groups continued as the most popular uses, however surfing on the internet had increased greatly. Latin American portals and chat rooms had become very popular.

A third study was carried out for the Inter-American Development Bank (IADB) in March 2000 (Proenza *et al.* 2000) based on 1,752 interviews in 14 *cabinas* throughout Peru, three of which were located in poor districts in Lima. The intention of this survey was to check whether the *cabinas* were, indeed, serving low-income groups. The results of this study have confirmed the profiles of users borne out by the previous studies: the users were young people (20.7 years old on average for students and 28.8 for non-students), who used the *cabinas* at least once a week for a period of one to two hours (in 68 percent of cases). Users visited an average of 2.3 *cabinas*, located less than 1 kilometer from their homes (in 44 percent of the cases). Most users were single (83 percent of the sample), and men outnumbered women (56 percent to 44 percent). Most had an education level above the Peruvian average (42 percent of them at university level). Forty percent had computers at home but only 5 percent were connected to the internet. In spite of the high education level, an analysis of family revenues showed that at least one in three users lived below the poverty line. With regard to the use of the *cabinas*, 57 percent of the students and 26 percent of the non-students interviewed said they used the internet for academic/school purposes. However, keeping in contact with friends and relatives was the main use for 24 percent of users. E-mail (60 percent), information gathering (51 percent), and chatting (39 percent) were the most frequent uses. Eleven percent of interviewees said they used *cabinas* to acquire computing and internet skills. Most users said that their computer skills had improved remarkably since they began visiting the *cabinas*.

These studies leave little doubt regarding the general profile of the users and the uses of *cabinas*. Regarding the users, the *cabinas* are serving the demand of a very young local public, with a high proportion of students and people with higher-than-average levels of education. Even though this profile applies more to the users of *cabinas* located in low-income areas, to some extent it can also be applied to users in other areas. Regarding the uses of *cabinas*, surveys and observations show that these have evolved over the past three years. Initially focused on work, school, and academic pursuits, sometimes unconnected to the internet, users gradually "discovered" the internet as a means of communication, resulting in the phenomenal increase in chat groups, e-mail, and surfing of the web. Later, voice transmission over the internet also became very popular and internet communications rapidly replaced a large percentage of traditional media such as mail and (national and international) long-distance telephony. Another change took place, mainly during 2001: the entertainment side of the internet was explored and adopted as a new popular application, especially by younger users.

The high proportion of young people (and students) visiting *cabinas* is clearly reflected in the type of activities engaged in. More than other age groups, young people favor chat groups, communication, and educational activities, entertainment and training for a future job. Overall, the proportion of users using computers for academic and training purposes remains high. This can be more clearly observed in the *cabinas* located in low-income neighborhoods, which attract a high percentage of students. John Zavalú (2001), owner of a *cabina* in Villa El Salvador, confirms that his business improves significantly when classes (in school and university) begin after the summer break. There is undoubtedly a strong correlation between education and internet use in Lima.[4]

A strong link with the business world is lacking, although activities such as job hunting and looking for business possibilities have been observed. More recently, different business initiatives have been launched by public and private institutions attracted by the possibilities of the *cabinas*. For example, the Commission for the promotion of small- and medium-sized enterprises (Prompyme) has established a network of *cabinas*, Cabipymes, to promote internet use by small- and medium-sized enterprises. In the same vein, the Tax Administration Service (SAT) has implemented a program, CabiSat, to promote the electronic payment of taxes through another network of *cabinas*. Terra Cabinas and EC-Cabs promote their businesses in a similar way. Governments have developed sites and gradually increased the variety of information, payments and other services, as well as means of citizens' participation which are available online.

Analysis of Users and Uses in Lower-Income Groups

In 2002, 57 percent of all internet users in Lima belonged to low- or very low-income groups (Apoyo 2002b; see Figure 7.3) (note that in 2000, the C and D sectors represented 82 percent of the population of the city). This constitutes an important achievement if one recalls the economic difficulties faced by the population in these sectors. However, lower-income users are, on average, less experienced with the internet, given that they only began to use it when the *cabinas* business was already flourishing. In 2002, 57 percent of users from middle-income groups (B) had already been using the internet for a period of two to four years; however, 63 percent of users from lower-income groups (D) had less than a year's experience (Apoyo 2002b).

In July 2000, A. y G. Asociados conducted 1,005 interviews in the C and D sectors for the purpose of gaining a better understanding of users and uses of the internet in these sectors. The survey focused on the central areas of the three Cones, the city center and the port area. Twenty-two percent of interviewees declared they were actively using the internet. The highest percentages of

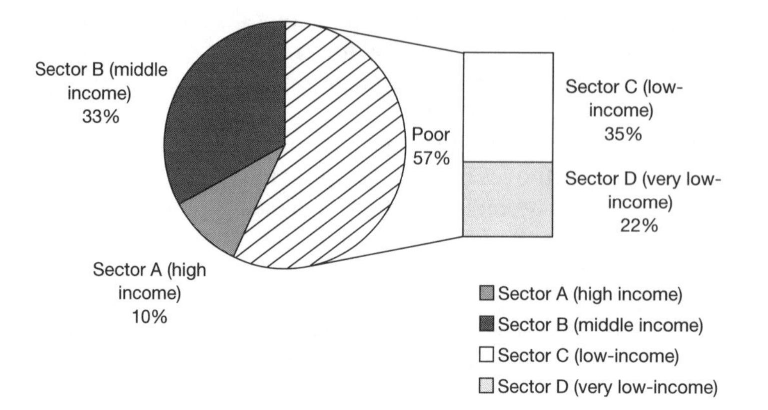

Figure 7.3 *Distribution of Internet Users by Socio-Economic Sector in Lima, June 2002*
Source: Data from Apoyo 2002b.

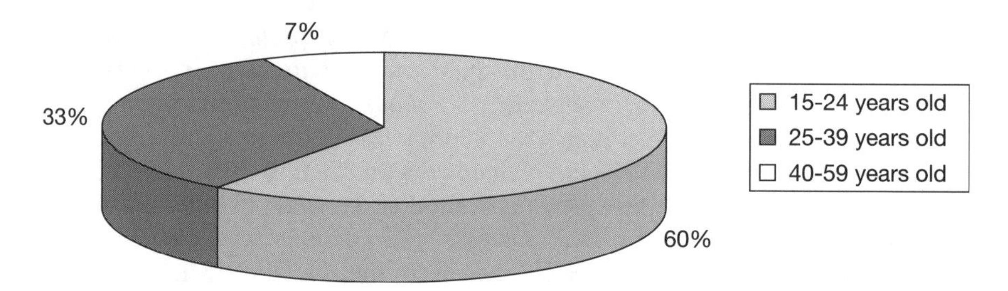

Figure 7.4 *Age Structure of Internet Users in a Sample Taken from C and D Sectors*
Source: Data from A y G Asociados 2000.

users were in the 15–24 age group (61 percent of men and 57 percent of women). The lowest percentage was found in the 40–59 age group from the D sector (less than 2 percent). The ratio of users between the C and D sectors was approximately 3 to 1 in all age groups. Figure 7.4 shows the age structure of the sample.

The proportion of men and women was almost the same in the younger and older age groups, however, in the middle group (25–39-year-olds), the proportion of male users was almost twice that of female users: 38 percent versus 20 percent in the C sector and 13 percent versus 7 percent in the D sector. These figures confirm the trend observed by field workers (Kerrigan 2001), whereby poor women in this age group have a triple function (their job, taking care of their children and households, and work for the community), which leaves them no free time for internet use. Only after their reproductive years do they show any interest in new technologies.

The survey showed that users from the C sector spent an average of 7.28 soles a week (US$2.10) on internet access in *cabinas*, while very low-income users (D sector) spent an average of 6.28 soles a week (US$1.80). These amounts represent 4.8 percent and 5.5 percent of average weekly household income in their respective sectors. This non-negligible percentage demonstrates the affinity of low-income users with the new technologies. The most popular applications in C and D sectors were e-mail, surfing the web and chatting, in that order, with a slightly higher preference in the D sector for chatting.

When asked about their main purpose in accessing the internet, users mentioned work and study as the two main reasons, followed by communication and social uses (email and chat). The main differences in the reasons given by low- and very low-income users concerned work-related purposes, which scored higher in the C sector, while chatting scored higher in the D sector. This can be explained by the higher presence of students in the D sector. Entertainment, reading the news and job hunting were also given as reasons for using the internet but scored much lower than the four main reasons. Figure 7.5 shows the percentages for the two income groups.

In terms of user activity profiles, students were by far the largest group in the low-income sector. This is hardly surprising among predominantly young users in a society where education is so highly prized. The percentage of students is even higher in the very low-income group. Another interesting feature is the percentage of retailers using the internet in both the C and D sectors (see Figure 7.6).

Another feature worth mentioning is the use of other media by internet users, which appears to be higher than average in the respective socio-economic groups, with the exception of movie attendance (see Figure 7.7). This leads us to conclude that as a group, internet users have a propensity to be "well

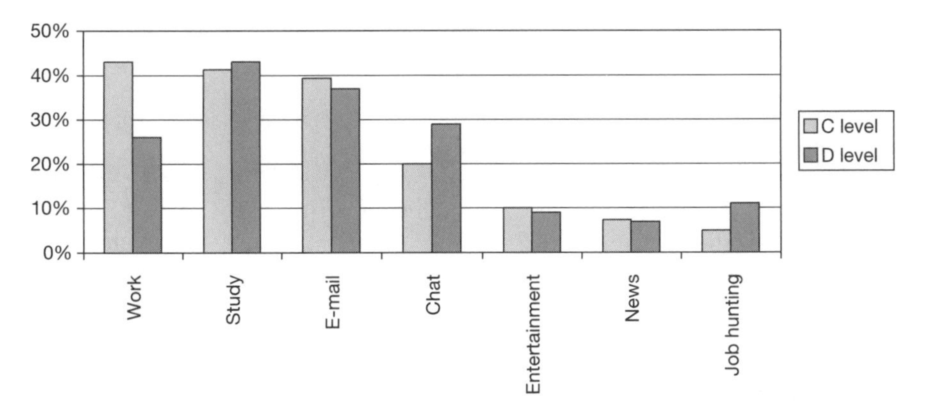

Figure 7.5 *Main Motivation for Using the Internet Among Low-Income (C and D) Groups in 2000.*

Source: Data from A y G Asociados 2000.

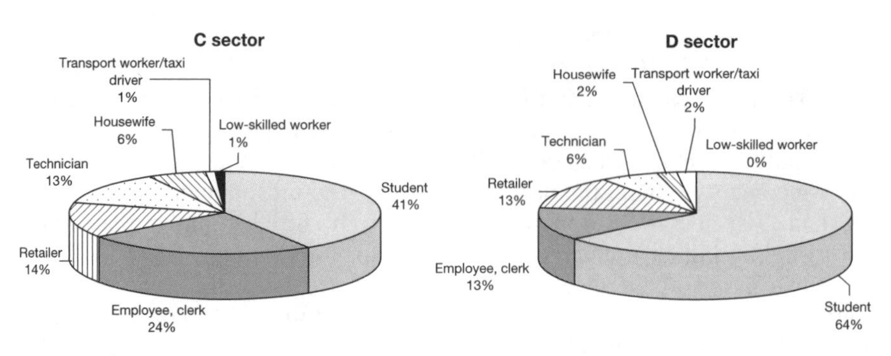

***Figure** 7.6 Job Categories Among Internet Users in C and D Groups*

Source: Data from A y G Asociados 2000.

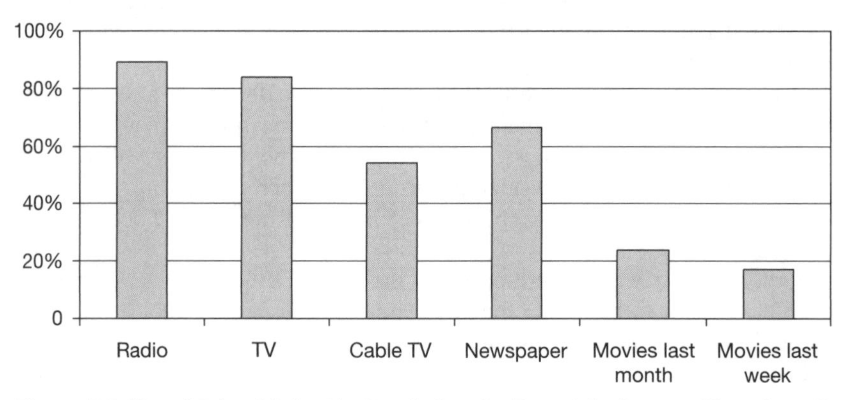

***Figure** 7.7 Use of Other Media (the Day before the Survey) by Internet Users from C and D Sectors*

Source: Data from A y G Asociados 2000.

informed." These results recall the correlation between newspaper penetration and internet penetration observed in Asian countries (Gray 2003).

In brief, the findings of this survey were useful in confirming general trends already identified in previous studies: Internet users include a high percentage of students, especially in lower-income groups; communication, work and study are the most important reasons for using the internet; also, an almost equal proportion of men and women use the internet. Other findings should be emphasized: low-income women in their reproductive years do not use the internet as much as women in other age groups, nor as much as men in the same age group; there is a strong link between education and internet use; internet users from low-income groups spend a proportionally large part of their income on accessing the net; low-income internet users tend to be better informed than average.

Cabinas, ICT Use, and Life Improvement: The Need for Institutional Steering

The effects of the presence of *cabinas* are widely visible in Lima's urban life, and not only because of their physical presence as a familiar urban facility at neighborhood level. Thus, the citizens of Lima enjoy relatively high levels of access to telecommunications with respect to the low levels of income of most residents. The high demand for telecommunications services is one of the main reasons for the successful emergence of *cabinas* and their sustained growth. What is it that makes low-income residents so eager to be connected, despite the high relative costs that this involves? It seems that being familiar with the internet and its applications has helped users to understand the potential benefits of ICTs. A brief description of the main transformations in urban life in low-income groups in relation to ICTs may provide clues to understanding their sustained interest.

Regarding education, the use of computers and internet by school and university students has become much more common. Teachers claim that the use of computers by students is gradually improving the quality of education. *Cabinas* are nourishing a new generation of ICT-literate people as children and the young are learning from each other how to use computers, and in some cases they are teaching their own parents. Furthermore, society definitely has a favorable attitude to the new technologies; no one questions the need to become part of the "information society" or the possibility of benefiting from ICTs. The internet is seen by most people as a medium for improving life chances and thus as something to strive to become familiar with. The idea that education and the acquisition of skills is the best way to achieve success in life is partly driving this huge demand for new technology as the results of the surveys confirm.

Regarding entertainment capabilities, Peruvians have changed their recreation habits and in 2002 visiting a *cabina* became their main recreational activity (InfoCab 2002). In fact, due to the multiple entertainment possibilities they offer, the *cabinas* have become an important part of the lifestyles of local youth. Newspaper articles report that street children and shoeshine boys are changing their habits and now go and play online games in the *cabinas*. Others give a colorful account of the relationship between the *cabinas* and popular culture (Jáuregui 2003). The topic has seized the imagination and interest of ordinary people.

In the socio-cultural life of the city the effects have also been pronounced. The *cabinas* have established their own place in popular culture. Peruvians appear to be proud of their *cabinas*, which are considered as something "typically" Peruvian (Jáuregui 2003) as the local media have emphasized the uniqueness of the *cabinas* phenomenon. The *cabinas* are clearly part of a cultural process related to the informal nature of the economy, which deserves further investigation (Venturo 2002). The topic has recently begun to attract the interest of local researchers and some initial interpretations have appeared. For example, the study of Colona (2003) on internet in everyday life has confirmed that, for young people visiting the *cabinas*, the internet represents an exciting new window onto the world, a link to mainstream global society (from which they are mostly excluded), and as such, the easiest way to "connect to modernity."

Other improvements in the living standards of poorer groups can be observed. Social networks, so important in the lives of the poor, are now extending beyond the locality. Indeed, since internet services are now more easily accessible, people have been much better connected with their relatives abroad than before,[5] a fact that has apparently helped increase the level of remittances to the country: Peruvians received US$1.26 billion from abroad in 2002, a considerable amount of money for such a poor country, which represented an increase of 24 percent on the previous year's remittances (MIF and IADB 2003).

Looking at the process at city level, interesting issues also emerge. In a city where the *barriadas*[6] constitute the main process of urban growth, there are major shortages of urban facilities at the neighborhood level. In this context and without any pretension, the *cabinas públicas de internet* are providing some important local facilities that had previously been absent in these neighborhoods. The libraries, recreation facilities, study places, youth centers, etc., that were traditionally lacking in the *barriadas* are now increasingly combined with the other services offered by *cabinas*.

Thus, it is not surprising that users declare that the internet is effectively improving their daily lives, especially those of young people and those from low-income areas. As we have argued above, there are also clear limitations regarding the gradual integration of ICTs into local economic development. However, an economy that relies heavily on the informal sector and cash transactions obviously faces huge difficulties in making the leap into the "network economy," especially given the quasi-absence of, and support from, local institutions.

The results of the introduction of the internet into society cannot be proclaimed beforehand and must be discovered through experience. This chapter has shown that in Lima, despite high socio-economic inequalities and great economic difficulties, and despite a lack of policies, programs, subsidies or support for promoting the development of ICTs, the *cabinas públicas de internet* have been highly successful in providing local internet access and services to less affluent groups. This reveals a high affinity and high demand from low-income groups for these new technologies that can be accounted for by the multiple advantages that internet access has brought about for lower-income users who are largely excluded from basic urban rights and services. Internet access is promoting significant transformations in terms of education, training, entertainment and communications that low-income groups in Lima have acknowledged as positive.

Thus, the experience regarding internet access in Lima has up to now been positive; however, questions about the future persist: will *cabinas* users be able to construct a convivial and democratic cyberspace or invent new ways of using ICT capabilities effectively as tools for local economic development? Internet access per se does not provide the answer to these questions. There are many possibilities for extending the benefits to other more sustainable goals while using the *cabinas* as an important starting point. In this process, leadership and support are indispensable: managing the leap from the informal economy to the network economy cannot be left to people who are living day-by-day, even if the creativity of the poor is immense. The role that the *cabinas* are fulfilling in the city, especially in the poorer neighborhoods, should be acknowledged, facilitated, and promoted. With the establishment of the *cabinas*, and in view of people's enthusiasm for visiting them, the first step towards the digital age has already been taken. It is now up to the institutions involved in Peru's development to assume their responsibilities and produce an integrated vision to guide the country's transition to the "digital future."

Notes

1 www.huascaran.gob.pe (accessed 17 October 2003).
2 This is valid only for premium telecommunications networks. Basic telephony (including internet service) is ensured, as the contract with the incumbent operator, Telefónica, obliges it to provide the service to any person that asks for it inside the city limits.
3 The sessions, contributions and conclusions are posted at: www.socinfo.concytec.gob.pe/foro_cabinas/default.htm (accessed 17 October 2003).
4 According to ITU officials, educational motivation and school enrollment are both strong predictors of internet use (Gray 2003). In both indicators Peru has a high score in terms of its income level (World Bank 2002).
5 It is estimated that approximately 10 percent of the Peruvian population lives outside the country.
6 Informal neighborhoods built in the periphery, generally by migrant residents.

References

A y G Asociados (2000) *Hábitos de Consumo del Usuario de Internet. Sectores C y D*, Lima: A y G Asociados. July.

Apoyo, Opinión y Mercado (2000a) *Usos y Actitudes Hacia Internet*, Lima: Apoyo, August.

—— (2000b) *Niveles Socioeconómicos de Lima Metropolitana*, Lima: Apoyo.

—— (2000c) *Perfiles Zonales de Lima Metropolitana*, Lima: Apoyo.

—— (2001) *Usos y Actitudes Hacia Internet*, Lima: Apoyo, October.

—— (2002a) *Usos y Actitudes Hacia Internet*, Lima: Apoyo, June.

—— (2002b) *Perfil del Internauta Limeño 2002*, Lima: Apoyo.

Colona, C. (2003) "Las Cabinas Públicas de Internet en Lima: Procesos de comunicación y formas de incorporación de la tecnología a la vida cotidiana," unpublished research report Pontificia Universidad Católica del Perú, January.

Drosdoff, D. (2002) "La Segunda Vida de Internet," *BIDAmerica. Revista del Banco Interamericano de Desarrollo*. Online. Available at: www.iadb.org/idbamerica/Spanish/APR02S/apr02s1.html (accessed 27 May 2002).

El Comercio y Apoyo Comunicaciones (2001) *Anuario 2000–2001*, Lima: Empresa Editora El Comercio.

Fernández-Maldonado, A.M. (1999) "Telecommunications in Lima: Networks for the networks?," paper presented at the Conference on Cities in the Global Information Society, Newcastle-upon-Tyne, November.

—— (2000) "Internet en Lima: un nouvel espoir?," *Les cahiers du numérique*, 1 (1): 235–45.
Gray, V. (2003) "Knowledge Indicators: measuring information societies in Asia-Pacific," paper presented at the Asia-Australasian Regional Conference of the International Telecommunications Society, Perth, Australia, June.
InfoCab (2002) "Hacia Buen Puerto," *InfoCab. Revista especializada para Cabinas de internet*, 1 (6), December.
Instituto Nacional de Estadística e Informática (INEI) (2000) *Tecnologías de Información y Comunicación en los Hogares de Lima Metropolitana*, Lima: INEI, August.
—— (2001) *Indicadores Tecnológicos de Información y Comunicaciones en los Hogares*, Lima: INEI, March.
International Telecommunications Union (ITU) (2000) *Americas Telecommunications Indicators*, Geneva: ITU.
Jáuregui, E. (2003) "Cabinas de la Choledad.com," *El Comercio* (Lima), 2 January.
Kerrigan, S. (2001) Interview by Ana Maria Fernández-Maldonado, March.
Lama, A. (1999) *Communications-Latin America: Internet Cheaper, More Accessible*, Inter Press Service. 29 December. Online. Available at: www.lib.nmsu.edu/subject/bord/laguia/lama.txt (accessed 21 October 2003).
Ludeña, W. (2001) "Lima: Poder, Centro y Centralidad. Del centro liberal al centro neoliberal," paper presented at the workshop "Las transformaciones de centralidad y la metodología de su investigación," Buenos Aires, November.
MIF (Multilateral Investment Fund) and IADB (Inter-American Development Bank) (2003) *Sending Money Home: An International Comparison of Remittance Markets*. Online. Available at: www.iadb.org/exr/prensa/images/RoundTablesFEB2003.pdf (accessed 21 October 2003).
Nagaro, F. (1999) "La Aparición de las Cabinas Públicas de Internet. Estudio de casos: Wilson y Villa El Salvador," unpublished research report, Universidad de Lima, Facultad de Ciencias de la Comunicación.
OSIPTEL (2003) *Indicadores del Sector*. Online. Available at: www.osiptel.gob.pe/ (accessed 10 October 2003).
Proenza, Francisco J., R. Bastidas-Buch, and G. Montero (2000) "Telecentros para el Desarrollo Socioeconómico y Rural," Documento de Trabajo del Departamento de Desarrollo Sostenible del Banco Interamericano de Desarrollo. Online. Available at: www.iadb.org/regions/itdev/TELECENTROS/ (accessed 28 September 2000).
Venturo, S. (2002) "Dilemas – Cabinas, Acceso y Redes Sociales," paper presented at the forum "Internet Cabinas: opportunities for all", December. Online. Available at: www.socinfo.concytec.gob.pe/foro_cabinas/default.htm (accessed 10 October 2003).
World Bank (2002) *Closing the Gap in Education and Technology*. Online. Available at: www.wbln0018.worldbank.org/LAC/LAC.nsf/ECADocbyUnid/A3CCD1D1859E48D185256CE5005F998B?Opendocument (accessed 10 October 2003).
Zavalú, J. (2001) Interview by Ana Maria Fernández-Maldonado, March.

CHAPTER EIGHT

Living in a Network Society: The Imperative to Connect

Sally Wyatt

Profiles and Paradoxes

The regular, biannual meeting of European Union (EU) heads of government held in March 2000 in Lisbon was dubbed the "dot com" summit, reflecting the realization by heads of state of the importance of information and communication technology (ICT) generally, and the internet in particular, for the well-being of Europe. The aim of the summit was to increase employment by promoting enterprise, competition and a dynamic, knowledge-based economy. To this end, the leaders agreed to reduce the cost of internet access to US levels within three years; to connect all schools to the internet by 2001 and to train teachers in its use (Tisdall 2000). From the heights of this summit, everyone was clearly understood to be a potential user of the internet. Access to the technology was seen as necessarily desirable and increasing access was the policy challenge to be met in order to realize the economic potential of the technology. Green and Harvey (1999) refer to this type of approach as the "connection imperative."

In the week of the meeting, the then Dutch Prime Minister, Wim Kok, admitted on national television that he had only recently attended his first internet training session (Nederlands 1, *NOS Journaal*, 23 March 2000). The British Prime Minister, Tony Blair, did the same in October 1999. As part of a campaign to promote ICT learning centers in poor parts of the country, Blair attended a two-hour training session in a shopping center where he learned about word-processing, email and the internet. He candidly admitted his ignorance of computing but asserted that the future of the country was dependent upon technological success (Smithers 1999: 8). Clearly, despite targeting the provision of ICT learning centers in deprived areas of the UK, internet have-nots are not only to be found among those groups of the population with low incomes and low educational achievements. Like the homeless, the British

Queen allegedly does not carry cash but she is perhaps more astute about the potential of ICT than her first minister. It was reported in March 2000 that she was set to become an internet millionaire following the success of her £100,000 investment in getmapping.com, a company which produces and markets complete, full-color digital maps of the UK (*The Independent* 20 March 2000). Since the dot com collapse in the latter part of 2000 however, her investment in late 2003 is worth only £30,000 (*The Guardian* 25 September 2003). The Queen is apparently a keen user of the internet but we can only wonder what web-sites she visits and if she uses an alias to surf the net in pursuit of interactions with the common folk. As an older woman without a university education, the Queen does not fit the profile of a stereotypical internet user.

There is a gap in the research about internet use and non-use. A literature search on various combinations of internet, computers, information technology, technology on the one hand, and rejection, dropout, non-use, barriers, have-nots on the other, yielded very few results. "Barriers" was the most productive, but much of that was about national-level adoption or education. "Dropouts" also provided quite a few references including some interesting material about young people who left school or university as a result of spending too much time online. The best title was the rather painful sounding, "Treating technophobia – a longitudinal evaluation of the computerphobia reduction program" (Rosen *et al.* 1993).

This chapter explores some of the paradoxes associated with the use of the internet and, more importantly, its non-use. The March 2000 European summit is only one of many examples of politicians and policy-makers assuming that access is the problem. From that perspective, making it cheaper and providing more education and training are among the obvious solutions. It is assumed that once these barriers to use are overcome, people will embrace the technology wholeheartedly. As Neice argues, "it is simply presumed by those advocating the elimination of the 'digital divide' that having internet access is always better than lacking it" (Neice 2002: 67). Access to the internet is seen as necessarily good. From the perspective of politicians, the hope is that people will then use this knowledge to create wealth and employment, but maybe they will use it to look at pornography, play games or trace long-lost friends and relatives. Maybe, some people will not use it at all and, difficult though it might be to accept, maybe its lack does not have to be a source of inequality and disadvantage. As Coutard (this volume) argues, "not all disparities among spaces in terms of the provision of [. . .] network infrastructures are socially undesirable" (p. 59).

The role of users has been increasingly addressed within the technology studies literature in recent years (Silverstone and Hirsch 1992; Lie and Sørensen 1996; Oudshoorn and Pinch 2003). In part, this reflects the recognition that users are not simply the passive recipients of technology but that they are active and important actors in shaping and negotiating its meanings. In part, the inclusion of users is an attempt to overcome the problems associated with those approaches that emphasize the powerful actors in producing technologies such as scientists, engineers, politicians and financiers. But focusing on use and consumption means that we are in danger of accepting the promises of technology and the capitalist relations of its production. Users of technology, often counter-posed to producers, also need to be seen in relation to another, even less visible group, namely non-users. For this reason, like Schneier-Madanes (this volume) I prefer the term "user" to that of "consumer." As she argues, consumers are most definitely bound up in capitalist relations whereas users can exist outside such relationships. In this chapter, I explore use and non-use of the internet. The second section reviews data about the profile of typical internet users. The third section presents the available data about people who do not use the internet, and explores what such data mean for policy. In the final section, I return to the issue of what the existence of large numbers of people who choose not to use the internet means for academic work about the network society.

Who Uses the Internet?

Since the development of the world wide web in the early 1990s, growth of the internet has been massive. Four host computers were connected in 1969, now the number is in the tens of millions. There are approximately 600 million individual users world-wide (Cyber Atlas 2003). Such dramatic growth tempts many commentators to conclude that this rate of growth will continue, or even accelerate. It is assumed the internet is following a very common path, one followed by many other successful technologies before it. Economists refer to this path as "trickle down"; the process whereby technologies that are initially expensive to use become cheaper over time, simultaneously providing more people with the benefits of the technology and enlarging the market. In the case of the internet, the early users were a small number of academics who used computers paid for largely from university budgets or defense contracts. Now, users include all sorts of academics as well as firms, political, and voluntary groups, and individuals at home.

According to the trickle-down view, there may be inequalities of access and use during the early stages of a technology but it is assumed these will disappear, or will at least be much reduced, as the technology becomes more widely diffused. Internet enthusiasts often claim that connection is a global process, albeit an uneven one. This is not unique to the internet. Similar claims can be found in much literature and in policy statements about industrialization and modernization more generally. However, as Lorrain (this volume) notes, such literature does not always distinguish sufficiently between the diffusion of goods and the diffusion of networks. The dynamics of network diffusion are more complex, contingent and dependent upon network externalities. Even so, individuals, regions and nations will "catch up"; those who are not connected now, will or should be soon. This is the real annihilation of space by time: the assumption that the entire world shares a single timeline of development, in which some groups are further along this path than others.

These views of trickling down or catching up contain several fallacies, two of which will be discussed here. (For a more extensive discussion, see Thomas and Wyatt 2000.) The first is that growth will lead to a more even distribution, whereas most of the available data suggest that it does not. Before presenting any data, a word of caution: collecting and interpreting data about internet use is not straightforward. Defining a host, ascertaining its location, identifying users and their demographic characteristics are all fraught with difficulty. Jordan (2001) demonstrates how estimates of the size and growth of the internet are frequently motivated by commercial needs and are not well informed by reliable sampling methods. With these caveats in mind, some patterns can be discerned.

Differences between countries remain stark. In mid-1998, industrialized countries, with less than 15 percent of the world's population, accounted for more than 88 percent of internet users. The US alone, with less than 5 percent of the world's people, has more than 28 percent of the world's internet users (Cyber Atlas 2003). This is broadly confirmed by Jordan's (2001) analysis of host computers. He finds that the world's richest nations (top 20 percent as measured by GDP per capita) are home to three-quarters of the world's host computers. Even within rich regions, differences remain. Over 40 percent of the populations of Sweden and the Netherlands are active internet users (defined as people who use the internet at least once a month); whereas less than 18 percent of French, Belgian, Italian and Spanish people are (Cyber Atlas 2003). (For a discussion of the different patterns of use in Peru, particularly the role of public access terminals, see Fernández-Maldonado this volume.)

The stereotypical user remains a young, white, university-educated man. However, closer examination of the available data indicates some deviation from this norm. Gender differences have shown the most dramatic reduction since the development of the world wide web, especially in the US. Georgia Technical University (1999) conducted online surveys of internet users approximately every six months between the beginning of 1994 and the end of 1998. In the first survey, only 5 percent of users were women. By October 1998, women represented just over one-third of users worldwide. By 2001, the Pew Internet Project (2001) finds that half of US internet users are women.

Differences based on race and income remain very marked. The first national survey in the US to collect data on ethnicity and internet use was conducted during December 1996/January 1997, based on nearly 6,000 respondents. Hoffman and Novak (1998) analyzed this data and found that whites are more likely to own home computers and to have used the internet than African Americans, even allowing for differences in education. The most worrying result occurs among high school and college students who do not have access to a home computer. Nearly 38 percent of white students, compared with only 16 percent of African American students, are nonetheless able to find some alternative means of accessing the internet. This may reflect different patterns of access within schools; or, the explanation favored by Hoffman and Novak, different schools may have variable levels of internet-related resources. (See McIver 2001 for historical analysis of African American access to ICTs.)

In *The Internet Galaxy* (2001), Castells draws on US data to support his argument that the digital divide is narrowing. He suggests that as the internet has been available for longer in the US than elsewhere in the world, it thus provides a good example of what will happen elsewhere in the world given time. Indeed, in Europe, policy-makers and analysts often look to the US to see what the future with the internet will look like, an example of applying the "trickle-down" perspective. The picture for the UK is not wholly dissimilar to that of the US. The 1997 British Household Panel Survey provides the most comprehensive data about the distribution of internet access from the home. Burrows *et al.* (2000) analyze this data and confirm that the typical British user is a young man, either a student or in employment, more likely to be living in London or the south of England. Not surprisingly, individuals with higher incomes and higher social classes are also more likely to be online. But Burrows and his colleagues do find some more intriguing results, very different from those found for the US. The most notable is that people from ethnic minorities were *more* likely to have home internet access than those who identified themselves as "white." Households inhabited by couples and children are more likely than other household types to have internet access. The gender gap in the UK and in Europe generally has not narrowed as it has done in the US. Approximately one-third of internet users in European countries are women (Cyber Atlas 2002).

The contours of the divide may vary between countries, reflecting national traditions of difference and exclusion, but social divisions in internet access exist, despite Castells' (2001) and Compaine's (2001) optimistic predictions that the digital divide is narrowing. The US Census Bureau conducted large-scale surveys (of approximately 48,000 households) on behalf of the National Telecommunications and Information Administration (NTIA) throughout the 1990s. The analysis of these surveys highlights what the authors call a "persisting digital divide." This is the same data drawn upon by Castells and Compaine, both of whom suggest the divide is narrowing. But the authors of the 2000 report, while noting that internet access has increased substantially, do not themselves support such optimism:

> *Nonetheless, a digital divide remains or has expanded slightly in some cases, even while internet access and computer ownership are rising rapidly for almost all groups*. For example, the August 2000 data show that noticeable divides still exist between those with different levels of income and education, different racial and ethnic groups, old and young, single and dual-parent families, and those with and without disabilities.
>
> (NTIA 2000: summary. Original italics)

A second fallacy implicit in the trickle-down assumption about continued growth is precisely that growth will indeed continue. The NOP Research Group (1999) has conducted regular large-scale surveys (of approximately 10,000 respondents) of aggregate internet use since the end of 1995. Between mid-1998 and mid-1999, they find that the number of internet users increased from almost 19 percent of the population to 27.5 percent, an increase of 46 percent. They predict the internet will continue to attract over 11,000 new users a day. But will this continue at the same rate until everyone is connected? Why should this be the case? Extrapolation is a notoriously unreliable forecasting technique, but one frequently used with respect to the internet. Paradoxically, the hopes often expressed by the promoters of new devices for accessing the internet, for example interactive televisions, palmtops and mobile phones, recognize that there is a point at which demand for internet use via personal computers will be saturated. Recent evidence of a flattening of internet growth in Europe and the US is presented in the next section.

Who Does Not Use the Internet? And Why?

The surveys referred to above are all concerned to demonstrate growth and, of course, growth has been phenomenal according to all available indicators, including numbers of hosts, domain names and users. Nearly all of the academic and policy literature focuses on how to increase the number of users, and takes the additional step of assuming that once a user, always a user. For example, the conclusions of Hoffman and Novak are to "ensure access and use will follow" (1998: 9). Moreover, they conclude, "programs that encourage home computer ownership . . . and the adoption of inexpensive devices that enable internet access over the television should be aggressively pursued, especially for African Americans" (1998: 9).

I will leave aside the question of indirect use of the internet (e.g. people making a query in a shop or agency where an employee uses the internet to provide the information needed). Instead, I will address the question of whether providing access is the sure, simple solution it sometimes appears to be. A recent report from the European Commission found that growth of internet penetration in homes leveled off during the second half of 2001, with the EU average being 38 percent (McMahon 2002). The reasons presented for this included the facts that people need computers in order to access the internet (this was presented as a startling new revelation) and the continued high (relative to the US) levels of connection costs. Cyber Dialogue (2000), an internet research consultancy based in the US, has also found evidence of a slowdown in internet growth, based on interviews with 1,000 users and 1,000 non-users. They claim that the rate of growth is slowing down overall and that there is evidence of an absolute decline in the number of users aged 18 to 29. In part, they attribute non-use to cost: as in Europe, some people cannot afford a computer and online access. They also claim that approximately one-third of all US adults simply do not believe they need the internet and what it offers. Even more significant is the growth in the number of adults who have tried the internet and then stopped using it; only one-third of whom expected they might use it again at some point in the future. In early 1997, they estimated there were 9.4 million former users; by September

1999, they calculated that there were as many as 27.7 million former users. Similarly, Winner (2000) discusses a survey of over 1,500 adults and 600 children, conducted on behalf of National Public Radio, the Kaiser Family Foundation and the Kennedy School of Government during November and December 1999. Winner does not provide the details but he claims that "a small but not insignificant minority" do not have a computer or any plans to acquire one. Three-quarters of this unspecified minority do not feel this as a lack.

Based on two national, random telephone surveys, Katz and Aspden (1998) suggest there are patterns to non-use. Their analysis of "internet dropouts" was a side effect of research about barriers to internet use in the US. They candidly admit they included the category of "former user" in their surveys only for logical completeness. They were surprised to discover in October 1995 that ex-users and current users each accounted for about 8 percent of the sample. They did another survey in November 1996, by which time the proportion of current users had more than doubled to 19 percent of the sample; the proportion of ex-users had also increased, but by less, to 11 percent. This survey took place prior to the Cyber Dialogue (2000) survey discussed in the preceding paragraph but it does provide more detail on reasons for non-use. People who stop using the internet are poorer and less well educated. People who are introduced to the internet via family and friends are more likely to drop out than those who are self-taught or who receive formal training at work or school. Teenagers are more likely to give up than people over 20. Older people are more likely to complain about costs and difficulties of usage whereas younger people are more likely to quit because of loss of access or lack of interest. Many ex-users have computers at home that they continue to use for other purposes, but not for internet access. This further complicates internet usage data because it is likely that many of these people will remain in the global statistics about ever-increasing numbers of users. More recently, the Pew Internet and American Life Project (Lenhart *et al.* 2003) has suggested on the basis of its regular surveys that the growth of the internet population in the US has flattened since late 2001, and is no longer showing the regular growth of the late 1990s.

As with all internet data, the Pew and Cyber Dialogue data and the results of Katz and Aspden need to be treated with caution as former users can, of course, become active users again at a later date. Nonetheless, the data are interesting because they call into question the assumption of never-ending growth. They also suggest that public access provision, quality of information and training remain important policy issues. If the results about teenagers are replicated elsewhere on a large scale, certain assumptions about the rate of exponential growth have to be re-examined. Maybe the internet is one of many things, such as sex and drugs, with which teenagers experiment only to abandon or use in moderation as they become older. Turkle (1995) draws on Erikson's theories of adolescent identity development to explain some of her observations of young people's behavior on Multi-User Domains (MUDs), the generic term for the huge variety of online, usually text-based role-playing games. Much early internet research, of which Turkle's is the best known, focused on such text-based applications. She suggests that MUDs provide a safe environment in which adolescents and young adults can experiment with different forms of interaction and relationships.

Other new technologies provide some pointers to patterns of use and non-use. Leung and Wei (1999) examine mobile phone use and non-use in Hong Kong. Mobile phones have a much longer history than the internet as a consumer technology. They identify the factors important in determining the take-up of mobile telephony. Age, income, gender and education all work in expected ways. However, age dominates: if you are older (unspecified), having more money and more education does not make much difference. Income levels are declining in significance, thus providing some support for the effectiveness of "trickle down." Mobile phones are no longer perceived as the preserve of smartly dressed, affluent, young men in suits. Intensity of use of mass media is not significant, but belonging to social groups that use mobile phones is. Equally unsurprising is the finding that non-users perceive the technology to be unnecessary because they have an alternative or because they find mobile phones complex to use (including pricing structures) or intrusive. Leung and Wei's results confirm a growing gap between communication rich and poor, with mobile phone users more likely to possess a range of alternative and complementary forms of telecommunication, such as pagers and answering machines; whereas non-users simply had one reasonable alternative. Leung and Wei accept the premise that having multiple communication devices is intrinsically good, whereas having only one adequate communication device is a sign of deprivation.

Leung and Wei's results are not very surprising: people do not use mobile phones if they have alternatives, find them intrusive and/or expensive. By extension, maybe some people do not use the internet because they have alternative sources of information and forms of communication, which are appropriate to their needs, or because they think it is cumbersome and expensive. As extrapolation, this is no more far-fetched than any other extrapolation found in discussions about the future of the internet.

The internet has changed significantly over the course of its existence, from an experimental network via academic resource to a general communication and transaction medium (for a fuller treatment of these changes, see Thomas and Wyatt 1999; see also Graham and Guy this volume). These changes have not stopped since the commercial "take-off" of the internet in the mid-1990s, as providers have developed and refined their services and strategies. How users perceive the internet, which to some extent determines how they use it, varies according to the time they first came into contact with it and the length of time they have been using it. It is possible that some of the explanation for why people might stop using the internet could lie in the difference between the time when the value of the internet was generally recognized, before or around the time of the "take-off," and the time when the majority of current users obtained access to it.

This could be translated into a potential gap between heightened expectations and the reality of the "internet experience." Internet hype can imply that the net is an easy way to access a cornucopia of exciting information, to communicate instantly and at low cost with people around the globe, and to buy a great variety of goods and services cheaply and conveniently. It can indeed be all of these things; but often it is, or it is perceived to be, something less than that. Since the commercial expansion of the internet, users have consistently expressed disappointment and worry over such problems as difficulty in finding relevant/wanted information and in navigating through the web, receiving unwanted information and keeping their personal details secure.

It is not difficult to imagine a cycle of cynicism and despair experienced by new users. At first, instead of the promised cheap and easy access to untold treasures, they receive more junk email than useful mail in their inbox and become frustrated with the waste of time (as well as money) while trying to access the information they require. They may become annoyed by the time taken up with the forced downloading of unwanted advertisements and with responding to inducements to click on links that promise more than they deliver. They may become bogged down in newsgroups where discussion of the notional topic is obscured by irrelevant and offensive contributions, or where the same topic is covered in multiple groups. They might also receive unwanted or inappropriate goods and services which were purchased online; and, then receive even more junk email as a result of information passed on to organizations as a consequence of some of these previous activities. The attractiveness of the "internet experience" for new users under such circumstances is likely to be severely diminished.

To what extent can such disappointing experiences be put down to the internet's temporary "growing pains?" The prognosis is mixed. Navigation tools are becoming more sophisticated, but so are marketing tactics designed to lure users towards information they may not have originally wanted. One way in which the complexity and chaos of the internet is being tackled is by guiding users towards a particular subset of "approved" content. This is the strategy of large-scale access providers, who design portals intended to provide "one-stop" access to facilities (with the emphasis being on facilities which can make money for the portal operators). In some cases, guidance is replaced by coercion, where users are restricted to accessing only a subsection of the internet approved by the access provider (and from which other users are excluded). While this may be an effective strategy for reducing complexity, it also reduces variety and choice. Patelis (2000) refers to this as a process of "e-mediation" in her critique of the America OnLine (AOL) portal. At the moment, it is not a major issue for most internet users because, in most industrialized countries at least, people have a wide choice of provider and can avoid such restrictions if they wish. It will become a much more important issue to the extent that reduction in the number of access providers restricts such choice.

Whether disappointment with the "internet experience" will lead to abandonment of internet use in part or in whole depends on a variety of factors such as: how much time and money has been invested in hardware and software purchases and learning to use the net; the availability of alternative means of accessing goods and services equivalent to those distributed over the internet. Another factor is how far the user is embedded in social circles that value and promote internet use. The declining amount of social prestige that can be gained from being an internet user as a result of the expansion of and the image of the internet promoted and reinforced by advertisements and by commentaries on the "e-commerce revolution" may also play a role. As noted elsewhere (Wyatt 2000; Flichy this volume), the metaphors used to describe and "sell" the internet have an impact on the way it is perceived, and hence on usage patterns. Reports of the internet as being primarily a vast reservoir of pornography may lead many people to delay internet access, especially within a family setting. Similarly, if the internet is primarily promoted as an e-commerce infrastructure ("online shopping mall"), then it is likely to attract different usage patterns from an internet that is presented as a universal information and communication resource ("library"). The commercial internet may be seen as tainted by individuals and groups opposed to the ethos of global capitalism, leading to avoidance or to attempts at resistance and subversion, depending on the predilections of the people involved.

The question of internet dropouts may only be a transient one if all dropouts eventually return to the internet, perhaps when their incomes rise or if they use one of the new access devices. Alternatively, the internet may follow the model of citizens band radio, the model of explosive growth followed by collapse. In any event, in the US alone, there are literally millions of former users about whom very little is known. They may be a source of important information for subsequent developments. Kline and Pinch (1996) vividly demonstrate the important role played by anti-car farmers in the US at the beginning of the twentieth century. Some rural inhabitants opposed the use of motorcars, and even after accepting its presence, used the car for a variety of purposes, such as grinding grain, plowing fields and transporting produce. Kline and Pinch demonstrate the significance of this for subsequent designs of motorcars and roads. Non-users might have something to contribute to design processes.

Even within the rhetoric of increasing access, it is important to know why some people stop using the internet. Internet service and content providers as well as policy-makers potentially have much to learn from this group. There are different categories of non-use. As Bauer (1995: 14–15) points out, there is a difference between passive "avoidance behavior" and active resistance. Also, care should be taken to distinguish between non-use of a technological system like the internet as a whole and non-use of specific services on it or aspects of it (Miles and Thomas 1995: 256–7). Some people might use email but never surf the web, for example. A preliminary taxonomy of non-use is presented below:

1 never used – because do not want to (resisters);

2 stopped using – voluntarily (boring, alternatives, cost, etc.) (rejecters);

3 never used – because cannot get access for a variety of reasons (excluded);

4 stopped using – involuntarily (cost, loss of institutional access, etc.) (expelled).

The policy implications are different for the different groups. For the first two groups it might be appropriate to develop new services to attract them. If internet access is seen as inherently desirable, this could be accompanied by the provision of measures to ease the transition from alternatives, as the French government did in order to encourage a switch from Minitel to the internet (OECD 1997: 25). Another possibility is to accept that some people will never use the internet. This could lead either to a focus on existing users or, moving away from the perspective of the suppliers and promoters who see non-use only as a deficiency that needs to be remedied, to policies that would ensure that alternatives to the internet were available to people who want or need them. The access issues identified at the March 2000 EU meeting related to cost, skill and location are more relevant for the third and fourth groups, those who would like access but who have been excluded for reasons beyond their control. In Marvin and Perry's (this volume) analysis of the strategies adopted by people who were forced to abandon their cars during the so-called fuel crisis in the UK in September 2000, they suggest another category of non-user, namely the "temporary" non-user whose short-term disconnection may lead to longer-term changes in strategy and behavior.

Only Connect?

Highways have been part of the discourse of widening access to the internet, especially during its "information superhighway" phase in the mid-1990s. However, many internet users almost immediately rejected Al Gore's superhighway metaphor, which projected the internet into public consciousness in 1994. Dyson *et al.* (1994) dismissed it as the worst possible description for the network society, largely because the engineering image of highways suggested to them a technology that was amenable to government control whereas they were keen to promote the image of an evolving, organic system.

The image of road kill on the information superhighway vividly represents the dangers of non-access, obsolescence and social exclusion (almost death) implicit in non-use. Elsewhere I have pursued the highway metaphor through a comparison of not driving a car and not using the internet (Wyatt 2003). Here I want to return to the methodological and theoretical problems of the "connection imperative." Acknowledging the existence of non-users accentuates certain methodological problems for analyzing socio-technical change. In the introduction, I highlighted the importance of incorporating users into technology studies as a way of avoiding the traps associated with following only the powerful actors. Another way of avoiding such traps is to take seriously non-users and former users as legitimate social actors who might influence the shape of the world. There are obvious methodological problems to be overcome here as non-users may be particularly difficult to locate, but nonetheless it may be worth the effort. In recent work (Wyatt *et al.* 2003), I have attempted to test the robustness of this taxonomy of non-users by examining the everyday experiences of internet users and non-users.

I shall conclude with a provocation: the use of the internet or ICT more generally by individuals, organizations and nations is taken as the norm and non-use is perceived as a sign of a deficiency to be remedied or as a need to be fulfilled. The assumption is that access to technology is necessarily desirable, and the question to be addressed is how to increase access. Sometimes the answer involves investment in infrastructure: public education to overcome ignorance and fear; training and standardization to improve ease of use. Informed, voluntary rejection of technology is not mentioned. This invisibility reflects the continued dominance of the virtues of technological progress, not only among policy-makers but also within the academic community itself. Castells claims that "exclusion from these networks is one of the most damaging forms of exclusion in our economy and in our culture" (2001: 3). Of course, I do not wish to condemn millions to economic or cultural exclusion, but it is important to remember that there are alternatives, that connection does not always have to be electronically mediated, and that non-use is not always negative.

Acknowledgments

The work on which this chapter is based was supported by the *Virtual Society? Programme* (www.virtualsociety.sbs.ox.ac.uk) of the UK Economic and Social Research Council, grant no. L132251050. I worked on this project with Tiziana Terranova and Graham Thomas, to whom I am grateful for many of the ideas developed in discussion with them. I appreciate very much the opportunity I had to discuss an earlier version of this chapter at the workshop, *The Social Sustainability of Technological Networks*, held in New York in April 2001. The discussion at the workshop and the subsequent written comments from the editors of this volume, Olivier Coutard, Richard E. Hanley and Rae Zimmerman, have helped me to develop the points raised here. None of these people share my responsibility for any of the mistakes in this chapter. I regret that my hard-won knowledge of water obtained during the workshop is not reflected here.

References

Bauer, M. (1995) "Resistance to New Technology and its Effects on Nuclear Power, Information Technology and Biotechnology," in M. Bauer (ed.) *Resistance to New Technology: Nuclear Power, Information Technology and Biotechnology*, Cambridge: Cambridge University Press, pp. 1–41.

Burrows, R., Nettleton, S., Pleace, N., Loader, B. and Muncer, S. (2000) "Virtual Community Care? Social Policy and the Emergence of Computer Mediated Social Support," *Information, Communication & Society*, 3, 1: 95–121.

Castells, M. (2001) *The Internet Galaxy*, Oxford: Oxford University Press.

Compaine, B. (ed.) (2001) *The Digital Divide, Facing a Crisis or Creating a Myth?* Cambridge, MA: MIT Press.

Cyber Atlas (2002) "Men Still Dominate Worldwide Internet Use," Online. Available at: http://www.cyberatlas.internet.com/big_picture/demographics/article/0,,5901_959421, 00.html (accessed 22 January 2004).

Cyber Atlas (2003) "Global Online Populations." Online. Available at: http://www.cyber atlas.internet.com/big_picture (accessed 23 September 2003).

Cyber Dialogue (2000) "Cyber Dialogue Study Shows Us Internet Audience Growth Slowing." Online. Available at: http://www.cyberdialogue.com (accessed 26 March 2004).

Dyson, E., Gilder, G., Keyworth, G. and Toffler, A. (1994) "Cyberspace and the American Dream: A Magna Carta for the Knowledge Age." Online. Available at: http://www.pff.org/position.html (accessed 20 May 2001).

Georgia Technical University (1999) *Internet User Surveys*. Online. Available at: http://www.cc.gatech.edu/gvu/user_surveys (accessed 26 March 2004).

Green, S. and Harvey, P. (1999) "Scaling Place and Networks: An Ethnography of ICT 'Innovation' in Manchester," paper presented at Internet and Ethnography conference, Hull, December. Online. Available at: http://www.les.man.ac.uk/sa/virtsoc/scale.html (accessed 20 April 2003).

Hoffman, D.L. and Novak, T.P. (1998) "Bridging the Digital Divide: The Impact of Race on Computer Access and Internet Use," working paper, Nashville: Vanderbilt University. Online. Available at: http://www.2000.ogsm.vanderbilt.edu (accessed 19 October 2003).

Jordan, T. (2001) "Measuring the Internet: Host Counts Versus Business Plans," *Information, Communication & Society*, 4, 1: 34–53.

Katz, J.E. and Aspden, P. (1998) "Internet Dropouts in the USA," *Telecommunications Policy*, 22, 4/5: 327–39.

Kline, R. and Pinch, T. (1996) "Users as Agents of Technological Change: The Social Construction of the Automobile in the Rural United States," *Technology & Culture*, 37, 4: 763–95.

Lenhart, A., Horrigan, J., Rainie, L. *et al.* (2003) "The Ever-Shifting Internet Population," Washington: Pew Internet & American Life Project. Online. Available at: http://www.pewinternet.org (accessed 16 April 2003).

Leung, L. and Wei, R. (1999) "Who are the Mobile Phone Have-nots? Influences and Consequences," *New Media & Society*, 1, 2: 209–26.

Lie, M. and Sørensen, K. (eds) (1996) *Making Technology Our Own? Domesticating Technology into Everyday Life*, Oslo: Scandinavian University Press.

McIver, W.J. Jr (2001) "Analog and Digital Divides: A Historical Examination of Access to and Discourses about Information and Communication Technologies in the African American Community," paper presented at the Annual Meeting of the Society for the History of Technology (SHOT), San José, California, October.

McMahon, T. (2002) "Internet Penetration in Europe Plateaus," Online. Available at: http://www.europemedia.net/shownews.asp?ArticleID=8308 (accessed 21 February 2004).

Miles, I. and Thomas, G. (1995) "User Resistance to New Interactive Media: Participants, Processes and Paradigms," in M. Bauer (ed.) *Resistance to New Technology*, Cambridge: Cambridge University Press, pp. 255–75.

National Telecommunication and Information Administration (NTIA) (2000) *Falling Through the Net: Toward Digital Inclusion*, Washington, DC: US Department of Commerce. Online. Available at: http://www.ntia.doc.gov/ntiahome/digitaldivide (accessed 23 April 2001).

Neice, D. (2002) "Cyberspace and Social Distinctions: Two Metaphors and a Theory," in R. Mansell (ed.) *Inside the Communication Revolution: Evolving Patterns of Social and Technical Interaction*, Oxford: Oxford University Press, pp. 55–84.

NOP Research Group (1999) "Internet User Profile Study, Wave 8 Core Data," Online. Available at: http://www.nopres.co.uk (accessed 26 March 2004).

Organization for Economic Co-operation and Development (OECD) (1997) *France's Experiences with the Minitel: Lessons for Electronic Commerce over the Internet*, DSTI/ICCP/IE(97)10/FINAL, Paris: OECD.

Oudshoorn, N. and Pinch, T. (eds) (2003) *How Users Matter: The Co-construction of Users and Technology*, Cambridge, MA: MIT Press.

Patelis, K. (2000) "E-Mediation by America Online," in R. Rogers (ed.) *Preferred Placement, Knowledge Politics on the Web*, Maastricht: Jan van Eyck Academie, pp. 49–63.

Pew Internet Project (2001) "Internet and American Life," Online. Available at: http://www.pewinternet.org/reports (accessed 16 May 2003).

Rosen, L.D., Sears, D. and Weil, M. (1993) "Treating Technophobia: A Longitudinal Evaluation of the Computerphobia Reduction Program," *Computers in Human Behavior*, 9: 27–50.

Silverstone, R. and Hirsch, E. (eds) (1992) *Consuming Technologies: Media and Information in Domestic Spaces*, London: Routledge.

Smithers, R. (1999) "Prime Minister Takes First Steps on Road to Email and Internet," *The Guardian*, 25 October, 8.

Thomas, G. and Wyatt, S. (1999) "Shaping Cyberspace – Interpreting and Transforming the Internet," *Research Policy*, 28: 681–98.

Thomas, G. and Wyatt, S. (2000) "Access is Not the Only Problem: Using and Controlling the Internet," in S. Wyatt, F. Henwood, N. Miller and P. Senker (eds) *Technology and In/equality: Questioning the Information Society*, London: Routledge, pp. 21–45.

Tisdall, S. (2000) "Will Europe's Third Way be the Dot.com Way?" *The Guardian*, 22 March. Online. Available at: http://www.newsunlimited.co.uk (accessed 26 March 2004).

Turkle, S. (1995) *Life on the Screen, Identity in the Age of the Internet*, New York: Simon & Schuster.

Winner, L. (2000) "Enthusiasm and Concern: Results of a New Technology Poll," *Tech Knowledge Revue*, 29 February. Online. Available at: http://www.oreilly.com/~stevet/net future (accessed 13 March 2004).

Wyatt, S. (2000) "Talking about the Future: Metaphors of the Internet," in N. Brown, B. Rappert and A. Webster (eds) *Contested Futures, A Sociology of Prospective Techno-science*, Aldershot: Ashgate, pp. 109–26.

Wyatt, S. (2003) "Non-users Also Matter: The Construction of Users and Non-users of the Internet," in N. Oudshoorn and T. Pinch (eds) *How Users Matter: The Co-construction of Users and Technology*, Cambridge, MA: MIT Press, pp. 67–79.

Wyatt, S., Henwood, F., Hart, A. and Smith, J. (2003) "De digitale tweedeling: Internet, gezondheidsinformatie en het dagelijks leven," *Amsterdams Sociologisch Tijdschrift*, 30: 254–73.

PART IV
Networks and Sustainable Access to Water

CHAPTER NINE

Conflicts and the Rise of Users' Participation in the Buenos Aires Water Supply Concession, 1993–2003

Graciela Schneier-Madanes

Introduction

The water concession of Buenos Aires,[1] Argentina's capital region, is currently the largest private water and sewerage concession in the world. Among the major cities that underwent privatization of their water services during the 1990s (these include Manila, Sydney, Jakarta, Mexico City, Santiago, Casablanca and Johannesburg), Buenos Aires' concession is a case reference on the subject because of the size of its population, the extent of the territory covered and the specific features of the privatization process.

This concession was part of a radical program of state reform and massive privatization encompassing virtually all public services and federally owned enterprises such as electricity, natural gas, telephone services, airlines, railways, subways, roads, ports and postal services initiated in 1989 in Argentina (Aspiazu 2003).[2] It took place within the context of a water supply crisis in Argentina and the internationalization of major European urban service providers. The reform was technically and financially supported by several international institutions (International Monetary Fund, World Bank). A regulatory agency, the *Ente Tripartito de Obras y Servicios Sanitarios* (ETOSS), was established for the regulation and control of water supply.[3] Its primary functions are to monitor the quality of service and to follow up contractual agreements. In principle, ETOSS is also responsible for determining rates. However, since the beginning of the contract, rates have been negotiated directly between the state and the company. In the central part of the urban region of Buenos Aires, a 30-year concession contract was granted by the national government to an international consortium called Aguas Argentinas (AASA) led by *Lyonnaise des Eaux*, now *Suez Environnement* (see Table 9.1). It started operating the Buenos Aires water system in

Table 9.1 Capital Stock Breakdown, Aguas Argentinas 1993–2000

Investor	*Capital Origin*	*1993 (%)*	*2000 (%)*
Suez Lyonnaise des Eaux-Dumez	France	25.4	34.70
Sociedad Comercial del Plata	Argentina	20.7	–
Sociedad General de Aguas de Barcelona	Spain	12.6	25.00
Meller	Argentina	10.8	–
Banco Galicia y Buenos Aires	Argentina	8.1	8.30
Compagnie Générale des Eaux (then Vivendi)	France	7.9	7.6
Anglian Water Plc	United Kingdom	4.5	4.30
Programa de Propiedad Participada	Workers	10.0	10.00
Corporacion Financiera Internacional	World Bank	–	5.00
Aguas Inversora*	Argentina	–	5.20

Source: Adapted from Aspiazu *et al.* (2002).
Note: *Meller economic groups.

1993. The concession's territory (2,000 km^2 with a current population of 9.6 million) consists of the city of Buenos Aires – the federal capital – and 13 municipalities (17 since 1998) adjacent to the capital and belonging to the province of Buenos Aires (see Figure 9.1 and Box 9.1), which are connected to the same water and sanitation system (or which are expected to be interconnected in the future).

For approximately 70 years after 1912, water and sewage management for the entire country was the responsibility of a state-owned company, *Obras Sanitarias de la Nación* (OSN). Following a decentralization reform in the early 1980s, the service area of OSN was reduced to the area that would subsequently become the concession's service area. Although OSN had achieved adequate service and coverage for a time, especially in the 1940s, the water supply and sanitation services were in a state of deep crisis by about 1980. As in most Latin American cities, the metropolitan area was expanding faster than the capacity of the waterworks, and networks in the city of Buenos Aires were in an especially bad state (Dupuy 1987; Rey 2001).

A short note on the origins of this crisis is useful here. Water was historically assigned a social function in Argentine society as one of the fundamental factors of hygiene and health. At the same time, there was a lack of awareness of its economic value and industrial dimension. Thus, one major principle in water politics was non-metered access to water, the so-called principle of *canilla libre* or the "free-tap" policy. OSN regarded water as an inexhaustible resource, available from the Rio de la Plata and underground water tables, and it had the final word on most water-related matters. Users and local figures (mayors or local administrators) simply had no say in this system.

By the end of the 1980s, the OSN was experiencing a series of problems, most of which are familiar to many water companies in Latin America (BID

Box 9.1 The Water and Sanitation Concession of Buenos Aires (2001)

City of Buenos Aires (CBA) and 17 municipalities of Gran Buenos Aires (GBA):

Area: 2,000 km^2

Population: 9,600,000 (2.9 million CBA + 6.6 M GBA)

Households below poverty line: 23.5 percent (Oct. 2001)

Households below indigence line: 7.4 percent (Oct. 2001)

Total clients: 2,625,000 clients

Billing: $554 millions (2001) (until December 2001 $1 = 1 *peso*)

Average water consumption: 600 liters per capita per day

Production: 4,155,000 m^3/day

Coverage:

Water: 81 percent

Sewerage: 63 percent

Sewerage treatment: 7 percent

Network data:

Water mains: 13,700 km

Sewerage: 8,600 km

Losses: 33 percent

55 percent of the infrastructure is over 60 years old

1997; Artana *et al.* 1999): in particular, only 73 percent of the population in the metropolitan area were connected to the water supply and 56 percent to the sewerage system (see Figures 9.1 and 9.2 for coverage rates in 1996–7). In suburban *barrios* not connected to networks (see Figure 9.3), residents obtain water from individual wells (with electrical or manual pumps) and sewage is disposed of through septic tanks or discarded directly into the ground, a system similar to those described in American cities in the early twentieth century (Tarr 1996).

When a connection to the wider network is possible but local infrastructure is missing, riparians or local communities frequently use the OPCT system

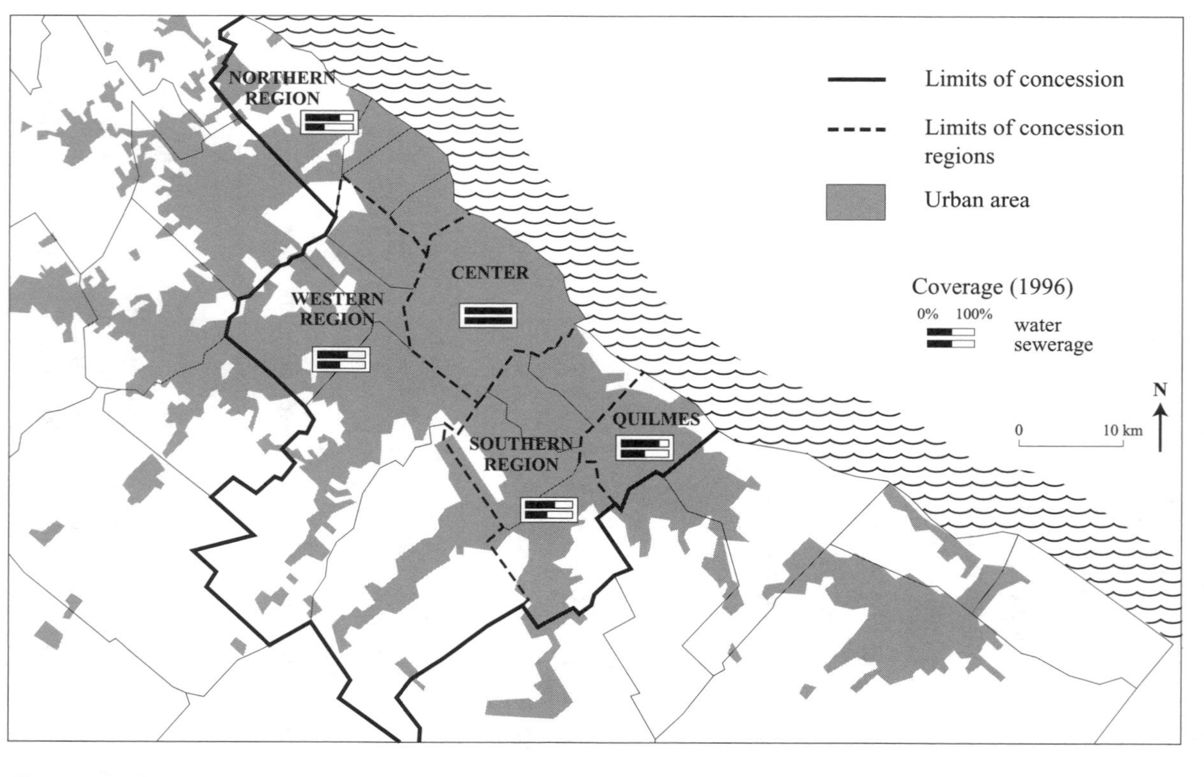

Figure 9.1 *The Aguas Argentinas Water Concession in the Buenos Metropolitan Area, with Indications of Coverage Rates by Water and Sewerage Networks (1996)*

Source: Data taken from Credal/CNRS (original data: Aguas Argentinas SA and ETOSS).

Note: Quilmes was integrated in the concession in 1998.

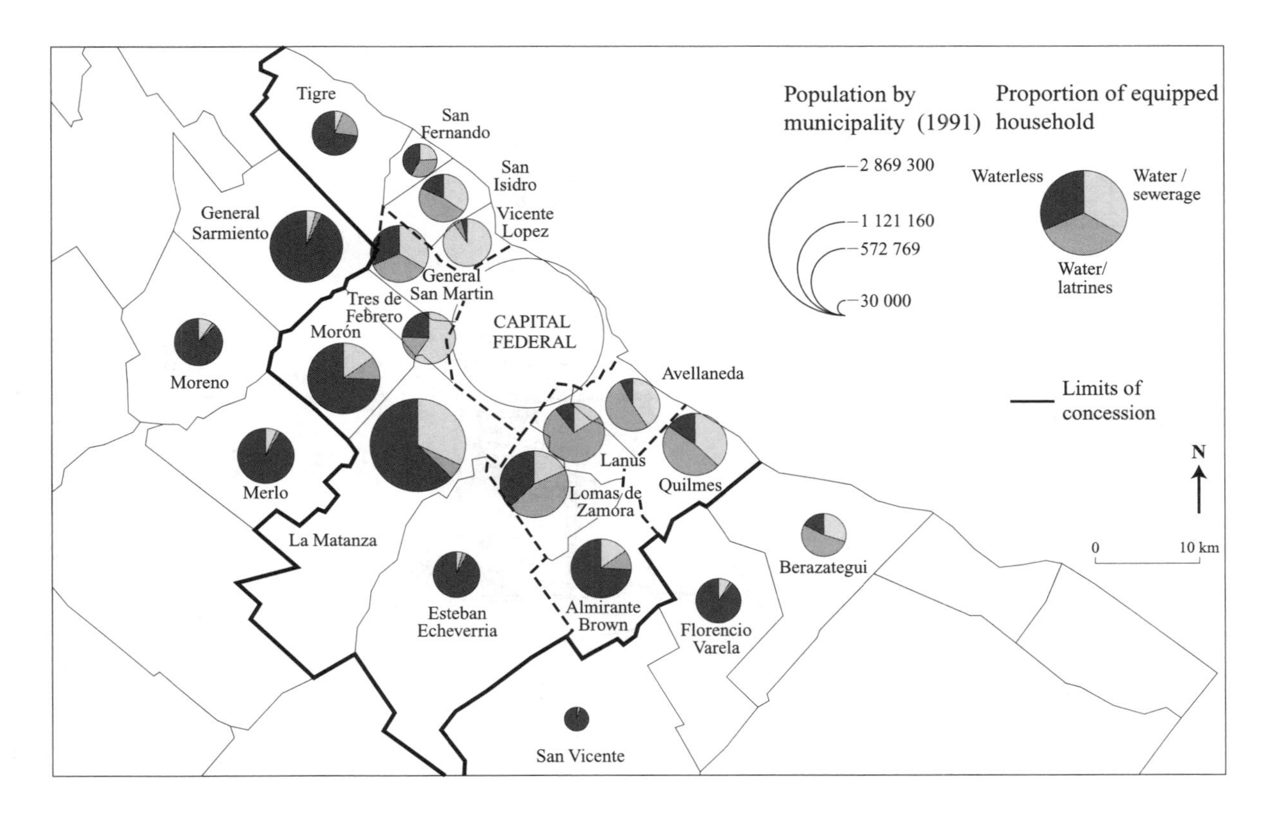

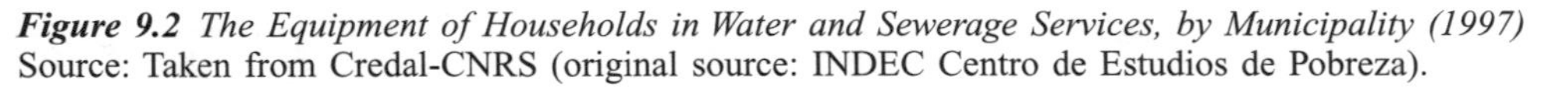

Figure 9.2 *The Equipment of Households in Water and Sewerage Services, by Municipality (1997)*
Source: Taken from Credal-CNRS (original source: INDEC Centro de Estudios de Pobreza).

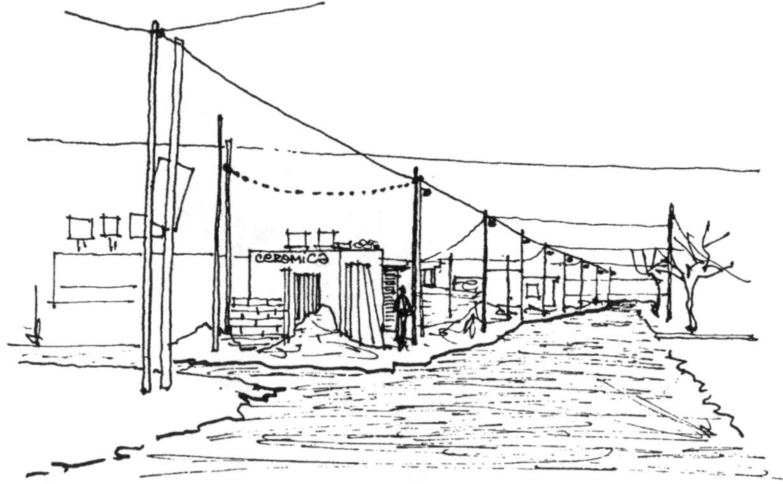

Figure 9.3 *A View of a Barrio.*

Source: Original drawings by author.

(*obras por cuenta de terceros*, i.e. work for third parties), in which they contract directly with public works companies to get infrastructures built. The OPCT system has existed since the 1950s for many services (street paving and lighting, electricity and gas supply, etc.). Although Aguas Argentinas was initially opposed to this system, it has been increasingly used in the water sector recently in reaction to delays in the implementation of the expansion plan. In many areas, cooperatives were created to develop local networks. In shanty towns people frequently had to obtain water from public access faucets, tank trucks or other legal or illegal sources (illegal connections were common).

The service concession was also designed in response to this failure of public supply (Dupré *et al.* 1998). The objectives of the concession were threefold: expansion of networks into previously unconnected zones, renovation of the existing infrastructure, and construction of sewage treatment plants. These objectives were part of a 30-year plan, broken down into five-year increments. At the end of the plan, water was expected to reach the entire population in the concession area and 90 percent of the population was to be connected to the sewerage system (see Figure 9.4).

This chapter examines the social and urban conflicts caused by the implementation of this plan. It is based on the assumption that major changes affecting water supply (construction, management, rates) have a social impact and transform not only the daily life of consumers, but also their relations with public institutions, how they perceive public utilities and, more broadly, their perception of water and the city in general. Conversely, these changes in perception also affect the development of public utilities. Conflict situations are particularly revealing in relation to such material, institutional and psychological changes.

Conflicts

The granting of the utility contract was a long process involving different commercial interests, including those of multinational corporations and international agencies. The contract was eventually awarded to Aguas Argentinas *SA* (AASA), which met all the relevant criteria and offered a rate 26.9 percent lower than that existing at the time of the bidding process.

Soon after the beginning of the contract, Aguas Argentinas requested, and ETOSS approved, a water rate increase of 13.5 percent and the contract was subsequently renegotiated on a regular basis. Mainly due to the introduction of new charges in addition to the basic water rate, the average water bill rose from $19.40 in 1993 to $27.40 in January 2002 (see Figure 9.5), i.e. from 8 percent to 11 percent of average household revenue (ETOSS 2001a). The minimum rate, which was $5 in 1997, rose by 60 percent in two years, to $8 in 1999, hitting low-income households particularly hard.[4] Following the first renegotiation of the contract in 1998, basic water rates were indexed automatically to the US consumer price index (Lentini 2003).

Within the "water arena" in Buenos Aires, increases in water bills created two successive conflicts: the first due to the introduction of a new charge, the infrastructure and connection charge (*cargo de infraestructura y conexión*, CIC) to be paid by all newly connected customers, and the second in response to the changeover from the CIC to a universal service and environmental charge (*cargo de servicio universal y medio ambiente*, SUMA) payable by all customers. These two conflicts are discussed in more detail below.

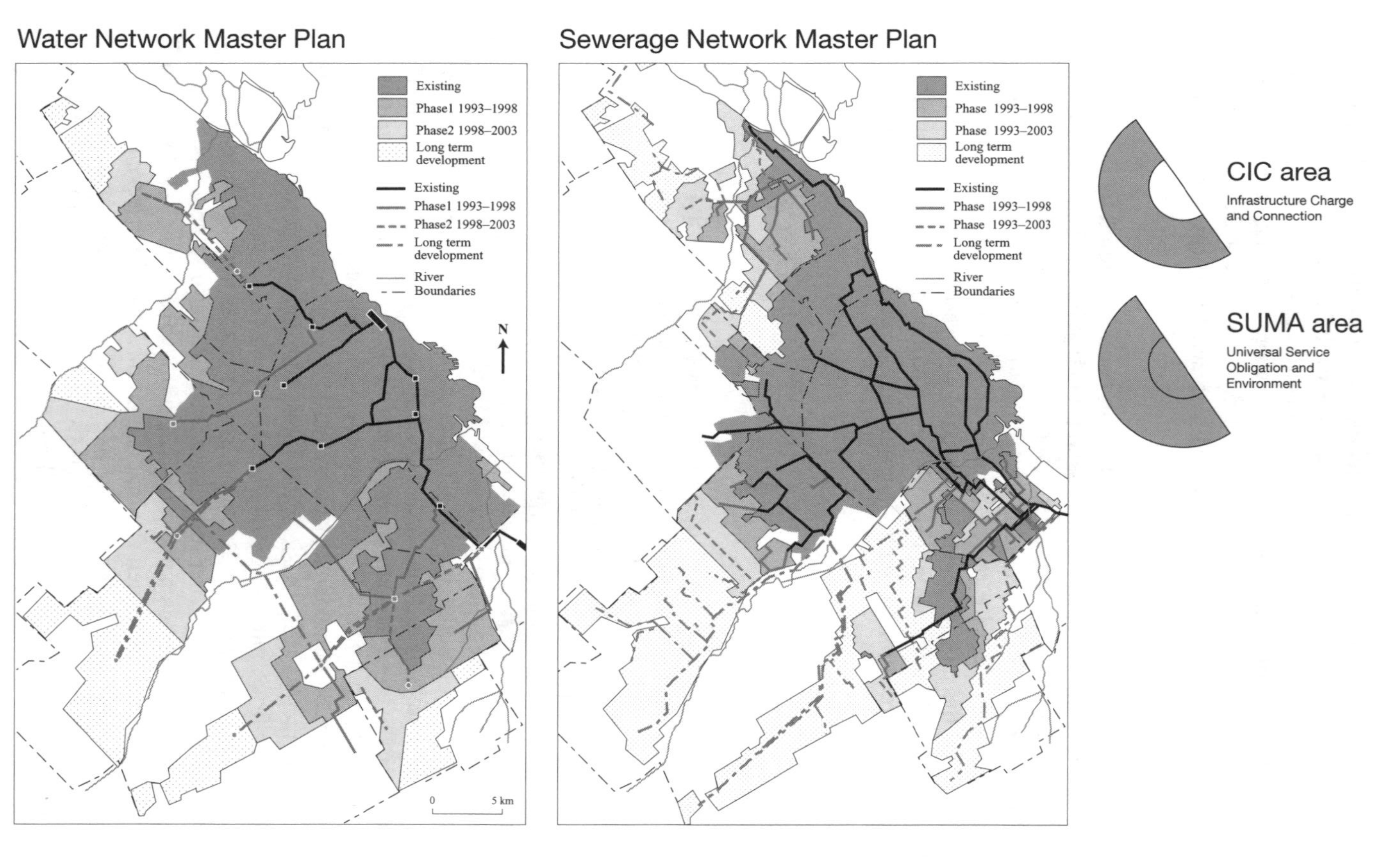

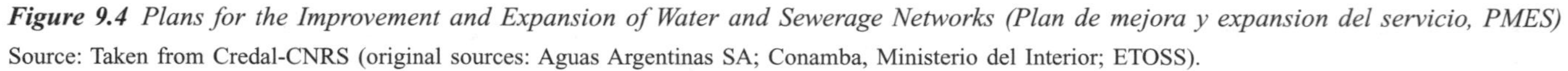

Figure 9.4 *Plans for the Improvement and Expansion of Water and Sewerage Networks (Plan de mejora y expansion del servicio, PMES)*

Source: Taken from Credal-CNRS (original sources: Aguas Argentinas SA; Conamba, Ministerio del Interior; ETOSS).

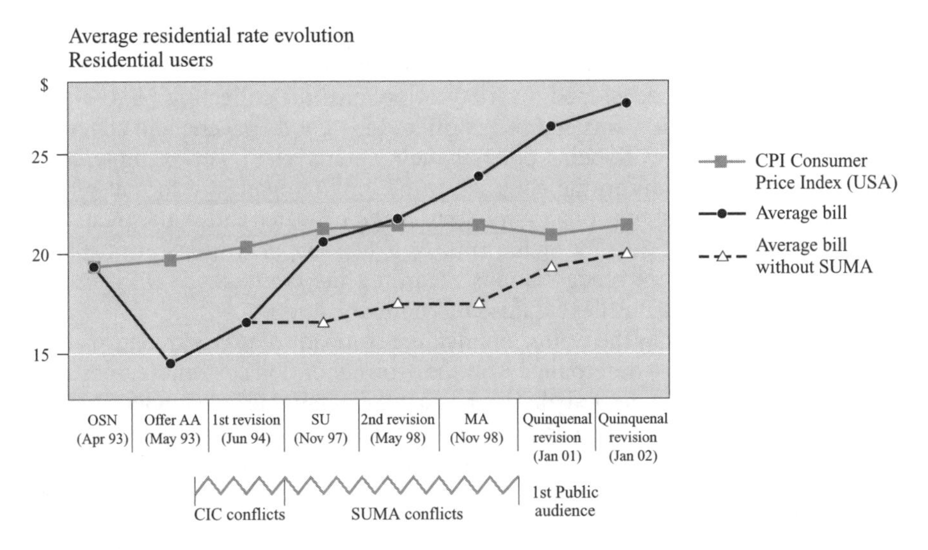

Figure 9.5 *Changes in Average Residential Rates and Conflicts over Water Services (1993–2002)*

Source: Data source for rates ETOSS.

Invoicing and Bill Collection

The pricing scheme partially inherited from OSN was a flat-rate scheme: the bill did not depend on the volume of water consumed but on characteristics of the property (size, location, building type, age, etc.). At the beginning of the 1990s only 5 percent of customers had a water meter, a proportion which subsequently rose to 13 percent.

Once they were granted the concession, Aguas Argentinas immediately set about updating the land survey and redefining its relationship with its two million residential, commercial and industrial customers. Customers received bills every two months and in the event of unpaid bills, the company would cut off the water supply after six months. However, from the beginning of the concession, the number of unpaid bills increased rapidly, due to regular increases in water rates and the gradual deterioration in the economic situation (Schneier-Madanes 1999). The impoverishment of a substantial part of the population of the metropolitan region became an obstacle to the implementation of the contract; in particular, plans to expand the network without public subsidies became unrealistic (Chisari and Estache 1999). Most full-income households can afford water and sewerage services. However, for the lowest income households, the cost of connection to these services was often too high due to the extra charges and, particularly, to the cost of home connection equipment which usually amounted to $1,500 (Villadeamigo 2003). Indeed, the proportion of low-income households increased dramatically as the national economy collapsed between 1998 and 2002.

Bill collection is a major problem for the company.[5] Over the first ten years of the concession, the company has continuously sought to sort "good" from "bad" payers and has developed a variety of policies for collecting bills. These policies include competitions and prizes offered to "good" payers, and advertising to show users the work being done on their water system. For example, bills may include ads relating to the contract or the environment; the format of the bill itself was changed to project a modern image of water and widespread use was made of advertising on TV, on the radio or in newspapers (Sinizergues 2003). Note that in the current renegotiations regarding the contract, ETOSS' users' commission questions this "excess use of communication."

In the early years of the water contract, the company also resorted to "social marking." Red crosses were painted on the front doors of customers with bad debt records. It was expected that exposing "disreputable" customers would result in pressure from those who paid for their water bills on those who did not. In fact, red crosses at times had the opposite effect; in some areas, neighbors continued to supply water to "bad payers" and prevented company staff from visiting the neighborhood. Other stigmatized customers reconnected their home illegally. As the economic crisis developed, these methods became very sensitive issues and were frequently debated in meetings and public hearings. Over time, the company abandoned some of these social engineering methods, replacing them with less aggressive ones, for example dispatching social workers to obtain community feedback information on the service or hiring former political activists to work with the residents and help to prevent such conflicts.

At the same time, the company developed tools to thwart illegal connections. An example is the "deep cut" (*corte profundo*) of the water and sewerage connections. When customers do not pay their water bills and the company decides to cut off their supply, it first resorts to a stopper, a simple device that can be removed easily and cheaply by the company – or by users wishing to cheat the company. For users regarded as particularly difficult bad payers, the company may resort to another form of disconnection, the "deep cut," which consists in dismantling the physical connection to the network. In contrast with the installation or removal of stoppers, deep cuts (and the restoration of the connection, if subsequently decided) are difficult and costly operations.

As a result of this combination of measures, the rate of bill recovery reached 95 percent in 2003 (according to ETOSS and AA staff), which the company regards as an acceptable rate.

Financing the Expansion of the Network: The Infrastructure and Connection Charge

Within such a context, rate increases could be expected to provoke opposition. The first conflict between the water company and users followed the introduction of the infrastructure and connection charge.

The contract specified that an infrastructure and connection charge (CIC) should be applied immediately to owners of newly connected properties. The charge ranged from $400 to $600 for water and $1,000 for sewerage, plus a connection fee, and it had to be paid in anticipation of future labor costs (Aspiazu and Forcinito 2003). Once the service was provided, the company would fill in all existing wells and septic tanks. Infrastructure charges could be paid over a period of two years in bi-monthly installments. Note that, since 1943, connection to the water and sewerage networks is compulsory for all residents living in connected areas.

Most low-income customers found it difficult to afford this new charge and a large proportion of them simply stopped paying their bills. Besides, many residents in the outskirts saw little benefit in connecting to the network as they already had ready access to ground water and had often spent money on drilling for water for their personal use.

Neighborhood protests began in 1995 in the western and southern low-income industrial municipalities (La Matanza, Lomas de Zamora) where local residents were able to compare the AASA rates with OPCT rates. Neighborhood associations, of which Villa Constructora (La Matanza municipality) is a symbolic example, prevented the continuation of projects and prevented the service provider from entering their neighborhoods. Thus, on one occasion, a human barrier of some 300 people stopped the engineering work on a project. In another instance, a tedious process of negotiation, mediated by ETOSS, complete with lawyers on both sides, was necessary, and led to changes in the project and its financing before it could be completed (Lacoste 1998; Schneier-Madanes 1999).

Resistance to the infrastructure charge was considerable and took various forms: formal complaints to the regulatory agency, street demonstrations, sit-ins in front of the company's regional headquarters, denunciations and presentations on television. Complaints to ETOSS and sometimes violent mobilizations provoked the closure of building sites and, on several occasions, compelled the company to stop working. Residents also frequently withheld bill payments in an effort to voice their discontent: the number of unpaid bills reached 80,000 in 1996 (according to various sources: ETOSS, AA).

Commenting on this conflict, Lorrain (this volume) rightly points to the fact that public utility companies do not enjoy a permanent position as suppliers: it is therefore risky for them to be involved in conflicts, either with their staff or with their customers. These companies also know they cannot operate a service in a city where a large part of the population remains excluded from this service for a considerable period. This can create an explosive situation.

Reforming the Infrastructure Charge: The Universal Service and Environmental Charge

In response to opposition to the infrastructure charge from those unconnected to the system, the government decided to change this charge to a "new universal service and environmental charge" (SUMA) to be paid by all customers (Lentini 2003). SUMA, in fact, consists of two separate charges: the universal service charge (SU) is aimed at covering the cost of network expansion, while the environmental change (MA) seeks to cover environmental investment (such as investment in sanitation facilities).

The billing formula for all non-metered clients thus became:

$$MF = TBB + (SUMA + CMC) \times FS$$

where: MF is the bi-monthly invoice amount (a lump sum, independent of the volume of water consumed); TBB is the basic rate, based mainly on the characteristics of dwellings; SUMA is the universal service and environmental charge (US$6 plus tax); CMC is the maintenance and renewal charge ($0.43); FS: service coefficient (water only: 1; sewerage only: 1; water + sewerage: 2).

SUMA is conceived as a mechanism for solidarity in that an additional charge on connected customers is expected to cover the cost of connecting new customers. In this manner, the burden of financing the expansion of the water supply and sewerage systems does not fall only on newly connected customers who usually belong to lower-income groups. The introduction of this charge meant a 13 percent increase in the previous residential invoice.

The conflicts generated by the introduction of SUMA (first the SU part, then the MA one) brought the water network into the political arena. As soon as the SU was introduced in November 1998, legal proceedings were instituted against ETOSS and the relevant ministerial department. The ombudsman acted, in particular, against the enforcement of the SU, emphasizing that the decision had ultimately been made by government officials and arguing that "if the SU is applied to all users of the utility contract, it should therefore be interpreted as a water poll tax." All year long, initiatives involving intimidation and conciliation alternated, while the controversy over the regulator intensified. Ministers, secretaries of state, political parties, members of parliament and associations all strove to influence the outcome of the process. "Water divides the country," one of the prominent national newspapers rightly pointed out (*La Nación*, 17 August 1998). SUMA also divided the metropolitan region (upstream and downstream) by opposing the city, which was demanding the renovation of its infrastructures and equipment for its urban projects, and the suburbs, to which the expansion of the network had become crucial.

The introduction of the SUMA charge had important consequences, especially with respect to solidarity among citizens in the concession area. First, it brought to light the existing inequalities in access to water and the contradictions surrounding the generalization of water and sewerage services. Second, it confirmed in the eyes of users the economic vulnerability of a concession that depended on their contributions. The conflict lasted until November 1998 when, following an agreement with the service provider, a presidential decree imposed the enforcement of the SUMA charge. Thus, the state did not assume responsibility for helping impoverished groups, but instead transferred this responsibility to the rest of the users. A year later the SUMA charge was integrated into the fixed part of the basic water rate and thus became "invisible" to users (Lentini 2003).

Social and Political Implications

The conflicts over rates and charges had a significant impact in terms of user and resident involvement in the regulation, organization and even the provision of water services. The constitution of 1994 had introduced obligatory user representation, but regulatory agencies, including ETOSS, ignored this requirement or resisted its implementation (Lopez and Felder 1997). The First Hearing on Water in December 1998 marks an important change in the administration of the water concession and that of privatized utility services in general.

Within the context created by the conflicts over utility services, two converging processes developed: the increasing importance of civil society organizations, NGOs and local governments, and a gradual recognition of the viewpoint of users by the government and regulators. Let us examine these processes in more detail.

The Institutionalization of Users

In the aftermath of the CIC conflict, new types of user organizations emerged. They had a strong link to local communities and can therefore be termed "locally based organizations." An example is the Users and Consumers Federation (*Comisión de enlace de usuarios y consumidores del conurbano*, CECUC), which was formed during the conflict in La Matanza. This neighborhood association, which initially grouped together around 50 residents, comprised both men and women from diverse political backgrounds, however, all of these people were concerned with water issues and, more generally, with urban life. As regards bill payment, CECUC and other locally based associations agreed on the necessity of paying bills, however, they asserted that the amount "must be fair" and they aimed at ensuring that this was so.

Local associations continued to develop in 2003 under various forms but they were not officially recognized. Their activities extended to other issues with a strong local dimension, e.g. rising water tables (3 million people are affected in Lanus, Lomas de Zamora, and other municipalities) and increasing water pollution (in Avellaneda or Quilmes). After the upheaval (*cacerolazo*: the term refers to the pans, or *cacerolas*, that people beat upon during demonstrations) in December 2001, which contributed to the fall of the Alianza government, a growing number of community groups sprung up and participated in intense social and political activity: regular and scheduled meetings, newspapers such as *Interacción Urbana* (which appeared in 1996 and defines itself as a "community-based local paper") or demonstrations like the "groundwater marches" (*marchas por las napas freáticas*) held against municipal and national authorities, and in front of ETOSS' offices in 2002 and 2003. Neighborhood-based actions multiplied across the political spectrum. During 2002, in a very uncertain political and social context, the most radical movements called for a *cabildo abierto* (a form of neighborhood committee that dates back to colonial times used in the fight for independence) in Morón, in the west of the concession area in order to terminate the contract; the movement of "self-summoned neighbors" (*vecinos autoconvocados* – i.e., a group which waited in vain to be consulted and which ultimately decided to "summon itself" and to take action) examines the issues relating to the privatization of the water supply; a large protest movement developed in the south of the area, especially in Quilmes, to protest against the rise in the water tables, for which local people blame the company.

These resident organizations are in keeping with a long social and urban tradition. Since the beginning of the twentieth century, urban growth in Buenos Aires has relied on community organizations (committees, cooperatives) involved in the development of local infrastructure (street paving and lighting, etc.). During the 1960s and 1970s, social movements in the periphery, organized on a local basis (neighborhood committees, shanty town associations), held protests in order to gain access to housing and land. Following persecution by the military dictatorship (1976–82), these organizations subsequently returned to the political arena. With the economic and social crisis of the 1990s, a weakening, or rather a fragmentation of these social networks occurred, and movements demanding basic subsistence rights appeared (Isla *et al.* 1999; Puex 2003), such as the *piqueteros* (from the French *piquet*, which means strike-picket), consisting of groups of jobless people who demonstrated regularly by blocking roads. The recent emergence of movements that refer to the "essential need" for water must thus be placed in the Argentinean political and social context.

Consumer associations gained official recognition under the constitutional reform of 1994 and the Consumer Protection Law (1998). These associations are traditionally made up of middle-class people and include a large proportion of professionals (lawyers, engineers, etc.). Their scope of activity has developed since the beginning of the privatization reforms. They seek to gain influence over service providers and, in particular, water companies. They did not get involved in the CIC conflict (which did not affect them), but actively opposed the SUMA charge. They advise users and consumers and act as a link with service providers. As regards the distinction between users (of public utility services) and consumers (of goods and services in general), it is worth noting that user associations consider themselves different from, and somehow more radical than, traditional consumer associations, in particular in terms of the former's claim that access to safe water and sewerage is a universal right. Moreover, the word consumer refers to goods while "user" relates to public utilities. Traditionally, there have not been any real differences between the two, however, recent conflicts have gradually made these differences more pronounced.

In 2001, 13 such associations were registered and are legally entitled to receive subsidies. With the exception of the experienced Consumer Action Group (*Acción del Consumidor*, ADELCO), they have all appeared fairly recently (less than five years ago). Several among them are linked to political parties (the Peronist, Radical or Socialist party) or to labor unions; they often emanate from cooperative movements (*Consumidores Libres*) or from certain areas of the city (Palermo Viejo, Belgrano "R"). One of the largest associations, the Users and Consumers Union (*Unión de Usuarios y Consumidores*, UUC), is a national organization. The power of these associations lies in their capacity to lobby Parliament or ETOSS. New consumer and user groups have emerged progressively, consolidating this new water rights movement.

The Participation of Users

The conflict concerning SUMA and the mobilization of consumer and user organizations forced ETOSS to convene, for the first time, a public water hearing. The hearing, which took place in December 1998, included representatives from professional and technical organizations, the company, consultants, workers trade unions, the media and, of course, officially registered consumer and user associations. (ETOSS 1998; Schneier-Madanes 1999). This list does not include a great number of unregistered associations (neighborhood committees, user federations), which also attended the hearing. A "users' commission" (*comisión de ususarios*) was created by ETOSS a few months later to serve as a kind of consultation group. In June 2000, a second hearing was organized to debate the "expansion plan" for the water network. On this occasion, the users' commission positioned itself as a defender of users' rights (ETOSS 2000).

The hearings appear to have been designed as a formal arena for participation – a kind of forum for holding discussions on the conflicts related to the water concession – rather than as arenas for problem-solving. They did not directly affect the decision-making processes. However, they provided considerable visibility on the issues involved in water supply, while arousing the interest of the public, elected officials and, of course, the media (Schneier-Madanes 2001).

The company and the regulatory agency agreed on one thing at least, namely the expertise of user associations: "the knowledge they have of the contract is startling. They are up to date on everything that is under legal consideration (contractual deadlines, rate increases, construction techniques . . .)" (Presidencia de la Nación 1998).

Simultaneously, the firm promoted a fundamental change in the perception of water by introducing the concept of "client/customer." In particular, it developed a sophisticated communications policy dealing with the new principles of water supply, the value of water and the need for avoiding waste, etc. (Sinizergues 2003). However, reactions to these initiatives were mixed. In general, residents question the notion of "client/customer" in a context where a single firm, Aguas Argentinas, holds a monopoly on water supply. In their view, a water user is best described as a "captive user" (*usuario cautivo*). This obviously constituted a limit to user participation.

NGOs and "Alternative" Solutions

Non-governmental organizations (NGOs) are becoming increasingly influential in water supply. This is in keeping with the tradition of social work in Latin America and Argentina since the 1960s, based on self-help movements, Catholic church groups and human rights associations, some of which are specialized in the question of water supply. NGOs seek acceptable paths for reforms, similar to those discussed by Barraqué (this volume).

Aguas Argentinas for its part, has contributed to social programs that emphasize local "alternatives" to the water network and which consist in the building of secondary water-only networks (no sewerage) by residents, or in setting-up collective organizations for mutual support. In some cases, municipalities supported these initiatives as part of the fight against unemployment. According to AA's department of sustainable development, ongoing programs in the southern part of the concession area mainly concern Santisima Trinidad (Quilmes municipality), Villa Besada, (Lanús municipality), Lealtad y Justicia (Avellaneda municipality) and, in the northern region, San Martín, La Paz, Perón, Evita, Antártida Argentina, Esperanza, San Cayetano (San Fernando municipality), Bajo Boulogne, Virrey Vértiz, Delfino, Cina-cina (Tigre municipality), La Cava chica, El Congo, Covicom (San Isidro municipality) (Aguas Argentinas 2003). Another example is the Riachuelo foundation. Since 1992 it has been conducting a pilot experiment in the shanty town of Villa Jardín (1,800 inhabitants) in the industrial municipality of Lanús (Lyonnaise des Eaux 1999; Schneier-Madanes and de Gouvello 2003). Generally speaking, the company has supported the emergence of a "grass roots level" in the organization of water supply. It should be noted though, that these experiments only concern a small part of the population within the concession area (around 10,000 people connected in 2003 with 200,000 more planned for 2003–5).

The Emergence of Local Communities

Local communities in the periphery were traditionally not in a position to assume direct responsibility for water and sewerage supply and they were not able to promote the integration of local demands into decisions because of their lack of representation in the regulatory agency. However, things have been changing recently due to the plan for improving and expanding the water and sewerage network (*Plan de Mejora y Expansión del Servicio,* PMES).

This plan lays down the guidelines for rate revisions and investment, as well as the corresponding technical and financial implications.[6] It is divided into five-year phases with intermediary expansion objectives. In line with the previous OSN approach, the expansion plan is based on technical and economic criteria, without taking into consideration the characteristics of the areas to be served, e.g. the administrative boundaries of the municipalities or the local demographic, social and economic differences between areas. For example, it does not take into consideration the fact that the areas most exposed to health risks are poor and densely populated, mostly located in the south of the metropolitan region, and it favors the extension of the network into the high-income north (Catenazzi 2003). It also ignores the "political climate" of the metropolitan region: national, provincial and municipal elections take place every four years.

The implementation of the plan was confronted with a diversity of local situations. First, mayors realized that there was an explicit water policy and wondered about the plan's local implications: when would the network arrive in their municipalities? What territories would be served first and how? For "managerial" mayors in the rich northern area, water supply was an uncomplicated policy area and their relationship with the service provider was based on mutual agreement. As for the "political" mayors in the western and southern areas, i.e., union-based mayors with strong, personalized relationships with their populations, they regarded public health and the social right to water as basic principles of their administration. This made them unconditional advocates of universal access to water and sewerage services. Communities in these places then began negotiating locally with the company over the numerous "adaptations" as to the timing and location of network extensions.

Second, starting in its regional centers in the different areas of the concession (north, west, south, city center), the company progressively gained a foothold at local level to become an important player in the municipal arena. Over and above its technical and commercial activities (network expansion and maintenance operations, billing, provision of client services, etc.), it became involved in local activities through social or cultural activities.

The ten-year period of local negotiations on water supply and sewerage services was a significant learning process and water expertise became a powerful political tool in these communities. Take the example of ETOSS' new social rate schemes (*programa de tarifa social*), which are aimed at poor families (ETOSS 2001b). These schemes are financed on AA's budget, with local mayors playing a key role by designating the beneficiaries of the scheme. The schemes were designed based on an innovative participatory process involving user associations, NGOs, the company and ETOSS.

The New Water Arena

The water concession of Buenos Aires reveals several aspects of the interaction between technical and social change.

The privatization reforms that affected public utility services in Buenos Aires took place in a social context hostile to publicly owned, public utility companies, due, in particular, to the combination of poor service and a financial crisis in many urban services, an unfortunate legacy of previous administrations. However, the population's initial support for privatization reforms was progressively undermined by rate increases in the majority of the services, the lack of subsidies to low-income families and the new commercial nature of the services. Such a situation was exacerbated by the impoverishment of large sections of the population within the broader context of an economic crisis.

Several conflicts and crises have tended to underscore major changes in users' perception of companies and regulatory agencies: the infrastructure charge (1996), the telephone rate adjustments (1997), the blackout in the summer of 1999 during which 200,000 people in Buenos Aires were without electricity for a period of up to ten days, the collapse of the electricity system in the winter of 2003 and, in September 2003, the (unannounced and thus doubly disruptive) one-day shut-off of the water supply in the city. There were common features in all these conflicts: the lack of a political authority to oversee them and the lack of independence of regulatory agencies. These factors fueled the severe criticism of privatization reforms. Several conflicts were subject to public hearings held following pressure from user associations.

Technical and economic changes (rate reforms, generalization of invoicing, introduction of new charges) gave rise to new social and public initiatives: the role of municipalities changed with the implementation of the water and sewerage network expansion plans. Municipalities became players in the regulatory process and arbitrators between utility companies and consumers. The two water conflicts (1996 and 1998) eventually led to the introduction and institutionalization of "water hearings" and the recognition of user committees as the official partners of regulatory agencies.

City-dwellers realized that a new bond connected them with the firm. This bond allowed them to be provided with water in their homes but, in exchange, they had to pay their bills and acknowledge that they had entered into a long-term relationship. The water bill symbolizes simultaneously the authority of the firm and the rights of the user. However, these new social relationships are hampered by the risk of disconnection and the increasing risk of marginalization faced by a significant part of the city's population. As in other Latin American cities (Fournier 2001), the current situation is radically different from the previous, now longed-for "golden age," where access to faucet water was regarded as a fundamental right that should be provided free of charge. In this sense a major change has occurred as regards access to water.

In addition, it can be argued that the company did not attain the objectives agreed upon in the contract. The long-term consequences of this situation (insufficient expansion, poor service, inadequate maintenance, etc.) may be serious in terms of the sustainability of the service. In this long-term perspective, a new scenario seems to be looming on the horizon. This encompasses the fight against poverty through the connection to networks, viewed as a fundamental objective which requires a change in how players perceive their way of life and which presupposes the transformation of such players into activists in modern urban society.

It should also be noted that privately owned utility companies, mostly controlled by foreign capital, have come to form a powerful lobby. As a result, the metropolitan arena has changed as these new powerful players increasingly intervene in the administration of essential services such as water provision. As their actions to some extent elude the control of urban decision makers, the organization of water supply in Buenos Aires, a key political issue, has thus taken on an international aspect (Schneier-Madanes 2003).

In the aftermath of the recent economic crisis, Argentina today finds itself in a very difficult situation. More than 20 percent of the working population is unemployed and more than half of the population now lives below the poverty line (Seoane 2003). The economic and social policies of the new government

(April 2003) have provided a basis for the discussion and renegotiation of privatized services and particularly the water concession of Buenos Aires. What will be the role of users and of civil society in general in the negotiations concerning public services? What lessons can be learned from the ten years of the Buenos Aires concession?

Obviously this new context gives rise to different and somewhat conflicting views of water facilities and of public utilities in general. Through the water crises, people became aware of underlying issues which eventually changed their perception of the city and its suburbs as the development of new settlements – poor as well as affluent – altered the urban fabric and living conditions in the entire city. The traditional opposition between the "center" and the "periphery" has become blurred; the increasing pollution and contamination of the water tables create new interdependencies and, possibly, new forms of solidarity; the functions and importance of old networks change, etc. These transformations entail new forms of management, and new expertise and innovative solutions, essentially because crises have network effects, and because, more often than not, their causes and consequences are not only local but also national and international. At stake are the issues and perspectives addressed at the Habitat Conferences (Vancouver, 1976; Istanbul, 1996; New York, 2001), and at the world summits from Rio (1992) to Johannesburg (2002), which dealt with the participation of populations in the production and management of their dwellings and living conditions, and sustainable development and environmental preoccupations in urban matters. Grass-roots organizations, NGOs and civil society, as well as international institutions are becoming increasingly important. In brief, water conflicts are clearly indicative of the state of the "urban and social question" in these times of increasing globalization.

Notes

1 The expression Buenos Aires refers here to the metropolitan region of Buenos Aires (12 million people in 1990), comprising the city of Buenos Aires (2.9 million), a self-governing entity and the capital of the Republic of Argentina, and a varying number of municipalities within the province of Buenos Aires (originally 19; 25 in 2003). The Buenos Aires water system has huge water distribution and sewage collection networks (some 11,000 km and 7,000 km of water and sewage mains, respectively), as well as an enormous water production capacity – 4 million cubic meters per day, of which more than 70 percent is produced by one treatment plant in Buenos Aires city. The main source of water supply is the River Plate and some 8 percent of water is supplied by deep wells located in peripheral municipalities. Most of the sewage collected (2.2 million cubic meters per day) is returned to the River Plate or flows directly into it without any treatment. Untreated domestic and industrial sewage flows into several rivers and creeks which flow through the metropolitan areas, and is discharged into the river.

2 Because of the need for large investments, especially for the expansion of the water supply and sewerage infrastructures, the concept adopted was the French concession model, whereby a private company (or a company with shared public and private ownership) assumes responsibility for operating, maintaining and investing in the system

over a long period (10 to 30 years, sometimes more), while the assets remain in public ownership.

3 The Board, which consists of six directors, represents the three jurisdictions forming the concession: the State, the Province of Buenos Aires and the city of Buenos Aires.

4 According to the definition of the Instituto nacional de estadisticas y censos (INDEC), low-income people are those barely able to afford basic foodstuffs. In 2003, the monthly income threshold was estimated at 710 pesos for a family of four. Indigent people are those who cannot even afford basic foodstuffs; as a reference, in 2003, 150 pesos were allocated to all jobless heads of households to help them cover vital expenses. (Furthermore, note that until January 2002 there was a fixed exchange rate 1 peso = US$1.)

5 This is revealed by the analysis of complaints received by ETOSS (i.e., complaints that were not satisfactorily resolved by the firm from a customer standpoint). More than 60 percent of the complaints received by ETOSS in 1996 concerned billing and the infrastructure charge, while more than half were concentrated in the six low-income municipalities in the southwestern area of the metropolitan region.

6 In 2000, water supply coverage was 80 percent (against a projected figure of 86 percent) and coverage for sewerage was 60.2 percent (against a projected figure of 78.6 percent).

References

Aguas Argentinas (2003) *Programa de barrios carenciados*, Buenos Aires: Aguas Argentinas.

Artana, D., Navajas, O. and Urbiztondo, S. (1999) "Governance and Regulation: a tale of two concessions in Argentina," in W. Savedoff and P. Spiller (eds) *Spilled Water: institutional commitment in the provision of water services*, Washington, DC: Inter-American Development Bank.

Aspiazu, D. (2003) *Las privatizaciones en la Argentina, Diagnosticos y propuestas para una mayor competitividad y equidad social*, Buenos Aires: CIEPP/OSDE.

Aspiazu, D. and Forcinito, K. (2003) *Historia de un fracaso: la privatizacion del sistema de agua y saneamiento en el àrea metropolitana de Buenos Aires*, working paper, Kyoto Water Forum.

Banque Interaméricaine de Développement (BID) (1997) *Privatisations des services en Amérique Latine*, Washington, DC: Banque Interaméricaine de Développement.

Catenazzi, Andrea (2003) "Universalidad y privatización de los servicios de saneamiento. El caso de la concesión de Obras Sanitarias de la Nación en la Region metropolitana de Buenos Aires 1993–2003," working paper, Buenos Aires: Universidad General Sarmiento.

Chisari, O. and Estache, A. (1999) *Universal Service Obligations in Utility Concession Contracts and the Needs of the Poor in Argentina's Privatizations*, policy research working paper, governance, regulation and finance, Washington, DC: The World Bank, November.

Dupré, E., Lentini E. and Chama, R. (1998) *Lineamientos regulatorios y tarifarios de los servicios de agua potable y desagües cloacales de Buenos Aires*, Buenos Aires: ETOSS.

Dupuy, G. (ed.) (1987) *La crise des réseaux d'infrastructure: le cas de Buenos Aires*, Paris: École nationale des ponts et chaussées (LATTS).

ETOSS (1998) *Audiencia Pública*, mimeo of the proceedings, Buenos Aires: ETOSS.

—— (2000) *Audiencia Pública*, mimeo of the proceedings, Buenos Aires: ETOSS.

—— (2001a) *Monthly reports*, Buenos Aires: ETOSS.

—— (2001b) *Programa de tarifa social*, Buenos Aires: ETOSS.

Fournier, J.M. (2001) *L'eau et la ville en Amérique Latine*, Caen: Presses Universitaires.

Isla, A., Lacarrieu, M. and Semby, H. (1999) *Parando la olla: transformaciones familiares, representaciones y valores en los tiempos de Menem*, Buenos Aires: FLACSO.

Lacoste, N. (1998) "San Francisco Solano", *Urbanisme* (n.s. Deux ports, un fleuve).

Lentini, E. (2003) "Diagnostic et solutions pour la régulation du service d'eau potable du grand Buenos Aires. Perspectives argentine et latino-américaine," in G. Schneier-Madanes, G. and B. de Gouvello (eds) *Eaux et réseaux, les défis de la mondialisation*, Paris: Travaux et Mémoires de l'IHEAL, La Documentation Française.
Lopez, A. and Felder, R. (1997) *Nuevas Relaciones entre el Estado y los Usuarios de Servicios Públicos en la Post-Privatización*, Buenos Aires: Instituto Nacional de la Administracion Pública (INAP), doc. no. 30, November.
Lyonnaise des Eaux (1999) *Solutions alternatives à l'approvisionnement en eau et à l'assainissement conventionnels dans les secteurs à faibles revenus*, Nanterre (Fr.): Lyonnaise des Eaux.
Presidencia de la Nación (1998) "*Los Servicios Públicos en la Post-privatizatión*," Buenos Aires: Jefatura de Ministros.
Puex, N. (2003) "Echanges, parenté, violence et organisation sociale d'une *villa miseria* du Grand Buenos Aires," PhD dissertation, Paris: IHEAL, University of Paris 3.
Rey, O. (2001) *El saneamiento en el àrea metropolitana, 1993–2000: Los primeros 7 anos de Aguas Argentinas*, Buenos Aires: Aguas Argentinas.
Schneier-Madanes, G. (1999) *Mondialisation des villes: les conflits de la concession de l'eau de Buenos Aires*, Paris: CREDAL.
—— (2001) "La construction des catégories du service public dans un pays émergent: les conflits de la concession de l'eau à Buenos Aires," *Flux*, no. 44–5.
—— (2003) "L'eau en jeu en Amérique latine," *Agir, revue générale de stratégie*, 16.
—— and de Gouvello, B. (eds) (2003) *Eaux et réseaux, les défis de la mondialisation*, Paris: Travaux et mémoires de l'IHEAL, La Documentation Française.
Seoane, M. (2003) *El saqueo de la Argentina*, Buenos Aires: Sudamericana.
Sinizergues, M. (2003) *Les images de la privatisation de l'eau à Buenos Aires*, Rapport Programme Ecos Sud "La gestion de l'eau en France et en Argentine," Paris: CREDAL.
Tarr, J.A. (1996) *The search for the ultimate sink: urban pollution in historical perspective*, Akron, OH: The University of Akron Press.
Villadeamigo, J. (2003) "Portrait économique de l'usager de l'eau à Buenos Aires," in G. Schneier-Madanes and B. de Gouvello (eds) (2003) *Eaux et réseaux, les défis de la mondialisation*, Paris: Travaux et mémoires de l'IHEAL, La Documentation Française.

CHAPTER TEN

Reforming the Municipal Water Supply Service in Delhi: Institutional and Organizational Issues

Marie Llorente

In India, as in many other developing countries, urban water supply and sanitation services are facing a major crisis. This crisis is aggravated by the pattern of urban growth: city dwellers will soon comprise one-third of the total Indian population,[1] as opposed to only 10 percent at Independence. There will be a real challenge over the coming years in bridging the gap between demand and supply, as cities constitute the "motor" of future economic growth. However, a poorly maintained infrastructure, intermittent supply, low pressure, water contamination, waste, leakage, budget deficits, and excessively low rates, etc. have been the norm in most cities. Consequently, the objective of achieving universal coverage, set out in the ninth five-year plan (1997–2002), appears to be very optimistic.

Although official figures indicate a reasonable level of coverage in 1991 (when 85 percent of the urban population had access to safe drinking water), these statistics do not reflect the true supply conditions implied by the poor operational performances of services. Moreover, there are strong disparities between states, cities and settlements. Compared with the national target of an average 140 liters of water per capita per day (lpcd), the real figures are significantly lower and range from 50 lpcd in most smaller towns to about 165 lpcd in a few larger towns (Suresh 1998). The lack of water particularly affects the urban poor: the volume of water available in slums is around 27 lpcd.

Since 1991, the new economic policy of the Government of India has been oriented toward liberalization and the potential role of the private sector is increasingly acknowledged. This is now clearly promoted in the water sector, both by the central government and financial institutions (the World Bank, for instance, is now well established in India). Indeed, there is a growing awareness and consensus about the need for substantial changes.

Nevertheless, the few attempts to initiate large-scale projects with private operators over the last few years did not turn out very well or have even been aborted (Mehta 1999). All of these projects consisted of supply-oriented partnerships, such as the construction and management of water treatment plants, and simply resulted in the addition of more capacity to a derelict network. On the whole, liberalization and partial privatization of the water sector did not bring the expected results. Low quality service delivery is endemic and in most cities the water crisis is becoming acute. Against expectations, the sector continues to under-perform and effective reform is still awaited.

The objective of this chapter is to understand why this sector appears so difficult to reform and why all efforts have failed up to now. The chapter consists of three parts. We first discuss why water as a resource and water supply as a service differ from other network utilities. This discussion will provide the framework for the case study. The chapter then focuses on water services in Delhi, and examines in detail the factors and stakes involved in the water crisis. Finally, we discuss the prerequisites for a sustainable reform of the sector, either under a public or private property rights scenario.

Water Supply Services Within an Institutional Perspective

From an economic viewpoint, urban water services display the archetypal characteristics of a network industry: substantial economies of scale (extending to natural monopolies), positive and negative externalities with regard to production and consumption and extensive vertical integration of activities. Yet, the remedies that have been tried elsewhere (in the gas, telecommunications and electricity sectors, etc.) – such as the unbundling of undertakings or the liberalization and/or privatization of supply, coupled with new regulatory mechanisms – have only rarely been applied to water and do not seem appropriate to this sector. The lack of activity regarding private arrangements in developed economies[2] and the moderate successes of the early experiments in developing countries suggest that the water industry has specific features that affect the scope of possible organizational forms.

Intrinsic Properties of Water and Their Institutional Implications

First, water has no substitute and is essential for life and health and for the process of economic development; it has cultural and religious significance for some populations, etc. For these reasons, it has traditionally been considered in many countries as an essential good and was often provided at a subsidized price or free of charge. This is particularly true in the rural areas of developing countries, where water is usually consumed directly at the source itself, and where the idea of paying for water when it becomes scarce is not easily accepted. In urban areas in such countries, where water is captured, transported (sometimes long distances), treated and supplied through a piped network, a charge is usually levied, however, it is under-priced, in the sense that the costs of production are not recovered. Water consumption thus remains largely subsidized and is often mismanaged, thereby generating waste.

Although this attitude is slowly changing, drinking water everywhere is considered a national priority.[3] As a corollary, water delivery as a basic service is also inextricably entwined with politics. Scarcity and inadequate supply accentuate political and administrative intervention at all levels. For example, at the national level, the Supreme Court of India has had to rule several times on interstate water disputes, however political arguments still occur; at the micro, local level, political leaders frequently maneuver for more public taps or water tankers in slum settlements. Patronage relationships are frequent in Delhi where slums' populations make up a significant number of voters (Haider 1993, 1997; Llorente 2002).

Second, water is a complex "system good," as it is both an ecosystem and a natural resource. The availability of water depends on the amount of withdrawals and on its very slow renewal rate. Besides, water is an incompressible resource, which means that it cannot be transported over long distances at a low cost. This is why water supply is organized on a local basis.[4] The local hydro-geological conditions are thus a key factor influencing the cost of water and, because water endowments are unequally distributed among and within states, this also explains why disparities occur so frequently and why this resource constitutes a strategic issue. That is why, in most countries, legislation provides for water property rights in cases of conflicting uses or interstate disputes. More generally, water resource management is now a necessity and it requires a well-defined national water policy translated into relevant actions.

Third, the water industry displays a high level of sunk costs (e.g. when investments are non-redeployable), since the pipe network accounts for a large part of total cost (up to 80 percent), a much higher ratio than in any other network industry. Consequently, the water sector offers fewer opportunities for competition among suppliers than other network industries such as gas, electricity or telecommunications. Unbundling in the water industry remains very rare, except for certain commercial transactions such as metering, invoicing and bill collection, which are sometimes franchised to private operators. However, water production and distribution are generally operated through a vertically integrated and publicly regulated "natural" monopoly. This governance structure predominates throughout the world.

Finally, water supply involves significant externalities,[5] in particular in relation to public health and the environment. If the quality of water is bad, its consumption will have a negative impact on health. Negative environmental externalities also appear when water is over-used or polluted. Environmental and quality controls are thus needed.

A Neo-institutional Approach to the Urban Water Industry

As the previous analysis shows, the organization and regulation of the water supply industry involves many aspects: economic, social, environmental, legal and administrative, political and ideological. In other words, one cannot understand the performance and functioning of this sector without considering the full institutional environment that determines the rules of the game. These rules are both formal (laws, policy, judiciary) and informal (customs, norms, codes of conduct) (North 1991: 97). In addition to these external parameters, there are of course internal factors within the governance structure (e.g. the public monopoly) that influence the performance of the sector: these include incentives, the degree of bureaucracy, the level of autonomy and skills of agents and the representation of public interests. Finally, individual behavior may also affect the governance structure, for example, if employees behave opportunistically (internal factor) or if users' associations play an active role (external factor) (see Figure 10.1).

Figure 10.1 reminds us that the governance structure of a water system does not operate in isolation. It is subject to macro and micro features that affect the performance of the system in different ways and to a different extent, depending on the local context. This means that reforming an urban water service cannot consist solely of a modification of the governance structure, e.g. by shifting property rights. This is all the more so when the institutional environment is complex and unstable.

New institutional economics provide an adequate conceptual framework for a broader-based study of the performance of water supply systems. Admittedly, industrial economics and new public economics provide interesting insights and prescriptions regarding the organizational structure and the regulation of network industries (Laffont and Tirole 1993). However, they focus on the governance structures of undertakings, the techno-economic characteristics of these industries (economies of scale, sunk costs, externalities, etc.) and the nature of bilateral relations between two agents (the firm – public or private – and the regulator). Furthermore, neo-classical approaches consider the institutional environment as an exogenous parameter, something that lies beyond the scope of economic analysis. This is a serious limitation when studying the performance of water supply, which we have shown to be highly sensitive to the uncertain and complex nature of the institutional environment. New institutional economics, by contrast, includes the institutional dimension as an endogenous parameter: this is why such an approach has been preferred in this chapter.

The Water Crisis in Delhi, its Origins and Consequences

Delhi is a very old city that has existed since the tenth century BC and 17 Delhis have so far come and gone in various locations. Today, it covers 1,483 square kilometers and overlooks the river Yamuna and the northern ridge (see Figure 10.2). Delhi has a special and complex

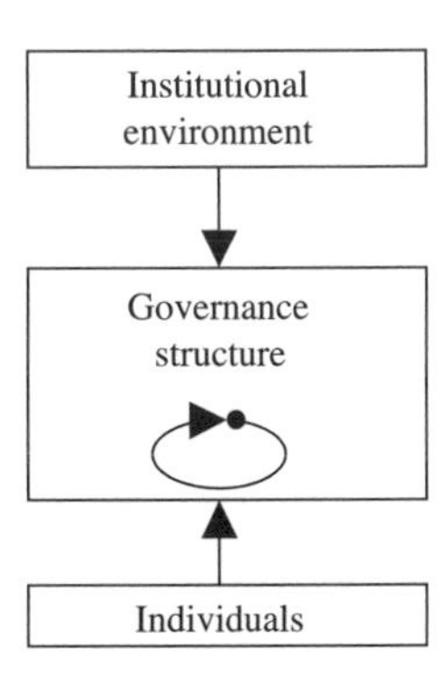

Figure 10.1 Water Supply and its Institutional Environment

Source: Adapted from Williamson (1993).

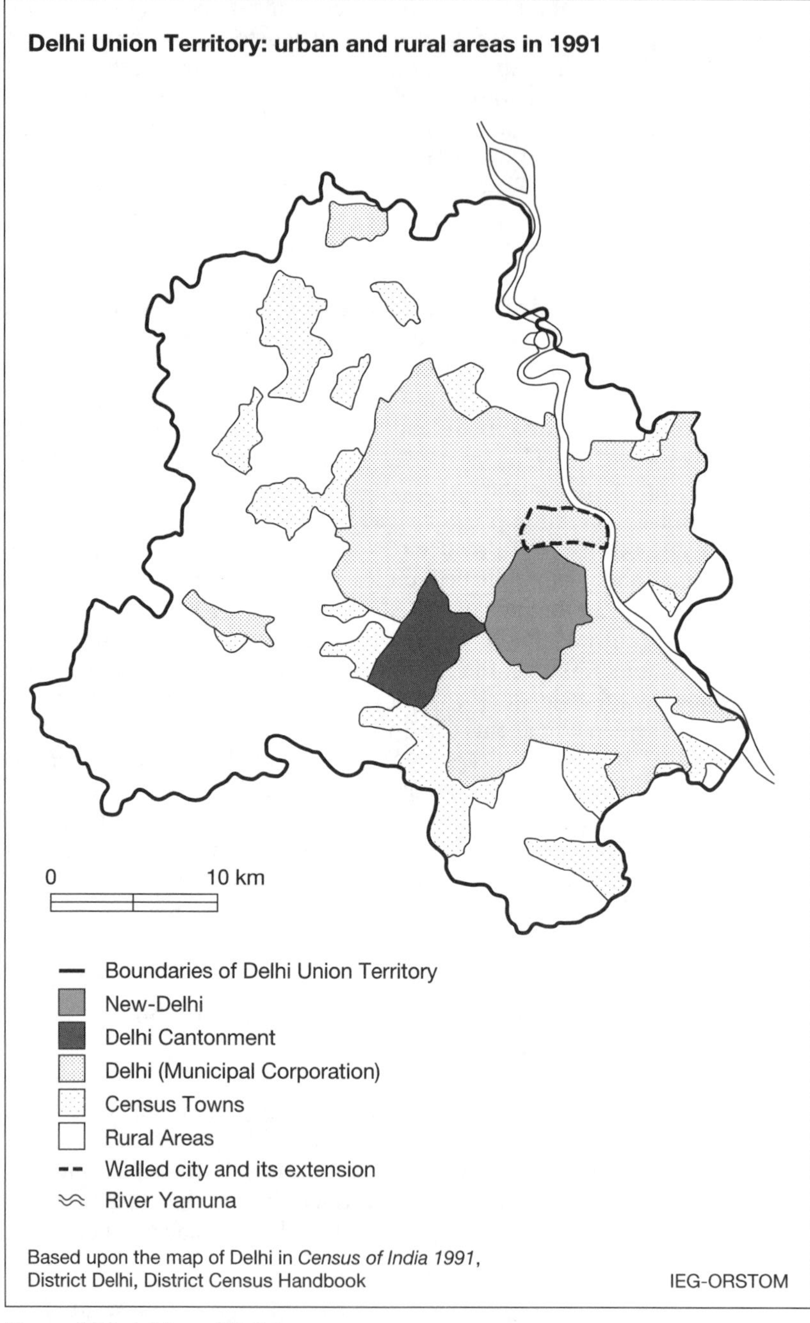

Figure 10.2 A Map of Delhi

Source: Taken from IEG-ORSTOM.

administrative status as the Union National Capital Territory: it is a State with a legislative assembly, although the supervision of Central Government remains significant. The Delhi Union Territory (DUT) comprises both rural and urban areas: 209 villages, 29 census towns and 3 local urban bodies: the Municipal Corporation of Delhi (MCD),[6] the New-Delhi Municipal Council (NDMC) and the Delhi Cantonment Board (DCB). The most salient feature of Delhi is its rapidly growing population, which now exceeds 10 million inhabitants; this raises serious problems concerning land occupation, habitat and access to basic services for all, as approximately 20 to 25 percent of Delhi's population lives in slum areas.

The Water Crisis in Delhi

In Delhi, the situation regarding water has long been a cause for alarm. Over the last 40 years, the continuing influx of people (500,000 to 600,000 per year) has generated a rapidly growing demand, which is now much greater than supply capacity. Every year, different parts of the city face water shortages and the problem becomes particularly acute in the summer. With the growing number of people arriving in the city and settling every year, the problem is likely to deteriorate.

The Delhi Jal Board (DJB) is the undertaking in charge of the provision and maintenance of water services. It was set up on 6 April 1998 through an act of the Delhi Legislative Assembly incorporating the previous Delhi water supply and sewage disposal undertaking. The DJB is responsible for the production and distribution of potable water following the treatment of raw water from various sources (including the river Yamuna, Bhakhra Storage, Upper Ganga Canal and Groundwater), while it is also in charge of the treatment and disposal of waste water. The DJB provides water in bulk to the NDMC and Cantonment areas. Sewage from these areas is also collected for treatment and disposal by the DJB.

Today, based on what is admittedly a very high, unrealistic consumption level (363 lpcd), water needs are estimated by the DJB at 3,993 million liters per day (mld) for an estimated population of 12.8 million, while supplies amount to approximately 2,613 mld. The gross availability of filtered water per capita is about 200 lpcd and only 150 lpcd when losses are taken into account[7] (CGWB 1996). There are huge contrasts between areas, with values ranging between 31 lpcd (in rural areas that are not always connected to the network) and 337 lpcd. Similar disparities exist with regard to water availability (which averages 3.5 hours per day).

Delhi relies mostly on surface water (the main source being the river Yamuna that flows through the city, although in summer its flow is limited). The other sources of surface water lie outside Delhi and water is shared through interstate agreements, however, these agreements are not always enforced as disputes sometimes arise during periods of crisis or scarcity. Moreover, there are significant transmission losses involved in bringing raw water from such distant sources. Regarding groundwater, the resource is already in a bad state: between 1977 and 1983, the water table fell by 4 meters in most parts of Delhi and this trend continued between 1983 and 1995, falling by as much as 8 meters in certain areas (CGWB 1996). Moreover, water depletion aggravates pollution due to a higher concentration of toxic substances such as fluorides.

As far as the resource is concerned, the city faces raw water availability problems mostly during summer, but it also suffers from pollution, which requires costly and elaborate treatment and, finally, property rights enforcement difficulties. The impact on health and the environment is extremely serious and poorest people are the most affected.

For slum dwellers,[8] the inadequate water supply results mostly in a time opportunity cost as they spend an average of three hours a day fetching water. In addition, they have to rely on multiple sources to ensure a sufficient supply for their families: 80 percent of households depend on at least two sources. The number of sources itself is very low (one source for 176 households) and far fewer than the standards defined by the program Urban Basic Services to the Poor (one public tap for 30 households or one hand-pump for 20 households). However, the most revealing statistic is the consumption of water, which is 27 liters per person per day. Moreover, it is estimated that, on average, these families spend between Rs 100 and Rs 135 per month in coping with the unreliability of supply. One should be careful before converting this amount into willingness to pay: that these households pay for water in an indirect way does not mean that they would accept having to pay "officially" for a better service.

Lack of financial resources is often used to justify the low quality of service, however, it is not really a satisfactory answer as the capital city receives a generous proportion of subsidies. Other factors must be investigated. The neo-institutional framework described above is useful for carrying out such investigations.

Insights on the Institutional Environment

According to North (1991), the institutional environment corresponds to the constraints that human beings impose upon themselves to structure human interaction so as to limit uncertainty. It consists of formal rules, informal standards and their enforcement characteristics. Understanding the way in which interactions are structured in the case of water supply thus requires an understanding of the related institutions.

As already mentioned, water is both an ecosystem and a resource. Some players have to deal with preservation issues while others are concerned with allocation problems so as to satisfy several competing needs (industrial, irrigation and domestic). In India, these conflicting objectives are all the more difficult to reconcile as there are at least seven levels of "water appropriation": the Union, water basin, state, region, city, local community and individual levels. Although, according to the National Water Policy Act, water is under the constitutional responsibility of the states, coordination between all these institutional levels is not an easy task since water is also a cross-sector resource. In the particular case of the supply of drinking water, several governmental institutions intervene regarding environmental and urban issues.

In order to gain a better understanding of the institutional framework, let us briefly present the major players, both at the central and state level. Central level institutions include the *Ministry of Water Resources*, responsible for laying down policy guidelines and programs for the development and regulation of the country's water resources; the *Ministry of the Environment and Forests*, responsible for environmental control and protection (in the case of water, this is mostly concerned with groundwater and water pollution) and represented at the state level through state pollution control boards; the *Ministry of Urban Affairs*, which is the highest urban development and housing authority of the Government of India at the national level and is responsible for formulating policies, sponsoring and supporting programs, coordinating the activities of various administrations and monitoring urban programs; and the *Planning Commission*, responsible for providing a national five-year plan and giving assistance to the states through the allocation of funds, sponsoring schemes, etc. These central institutions provide the framework for project design and implementation by laying down policy guidelines and regulations, and by transferring funds.

They are complemented (in the case of Delhi) by state level institutions: the *Delhi Jal Board*, the *Delhi Pollution Control Board*, attached to the Ministry of the Environment and Forests, and several *urban planning authorities*, especially the Delhi Development Authority and the National Capital Region Planning Board.[9]

As this breakdown shows, Central Government has set up a dedicated organization for each single water issue; not surprisingly, these organizations operate in their respective fields, "following their own perceptions and policies, and the department that has the most influence in government gets its way by getting approval for its projects, to the detriment of a well-coordinated water management policy" (Sinha 1995). Numerous players with overlapping functions and different powers and resources are involved in the sector. This intricate administrative environment contributes to the general confusion because of the proliferation of inadequate guidelines, resulting in inefficient planning.

These technical departments are also criticized for functioning in isolation from realities on the ground and for their lack of pragmatism (Narang 1997). For example, the numerous conventional master plans, development plans and the like, do not even acknowledge the existence of the informal sector, although it is a major provider of services.

The influence of town planners is also criticized for having imposed a western conception of the city, based on unrealistic standards of population density in view of the realities of the demand for land. As a result, in some cases "ghettos" have been created which stand in sharp contrast to their immediate surroundings (as is the case in the residential areas of New Delhi); in other instances, residential areas have been "invaded" by unplanned and illegal constructions which nevertheless correspond to local requirements and are the consequence of the pressure on vacant land (Milbert 1998). Finally, planning efforts are uncoordinated, disconnected from reality on the ground and mostly biased in favor of the middle and upper classes. This consequence is an accentuation of spatial segregation in the city according to level of income.

A look at the internal structures of these public agencies and public utilities provides some insights.[10] They all have a bureaucratic organization, based on a well-established hierarchy, which has long characterized Indian public administration. However, such a structure has several negative side-effects, such as significant delays in decision-making, problems with day-to-day management, inefficient planning, corruption and dilution of responsibilities, etc. Other parameters increase uncertainty and discontinuity, such as the turnover of administrative personnel (especially the highest-ranking municipal officials) who rapidly move from one position to another, or the lack of training of local personnel in order to enable them to adapt to changes. Within these organizations, information is poorly disseminated because of archaic procedures, poor resources, time-consuming paperwork, incomplete reporting, and lack of control. Politicians frequently use their external discretionary power, as can be seen in regard to tariff setting procedures and investment decisions. Last, the problem of coordination is reinforced by a "competition virus between politicians and administrators," as mentioned in a report of the national commission on urbanization (Ministry of Urban Development 1988).This fragmented and incomplete institutional environment is not credible enough to create reliable incentives for a substantial reform of the water sector and the option of independent regulation appears unrealistic in such a politicized and bureaucratic system, mainly because interference would persist.

Insights into the Governance Structure

Regarding the public undertaking itself, the problems to be solved are quite similar to those in other developing countries faced with high population growth. The so-called "vicious circle" perfectly applies to the context of Delhi (World Bank 1998): poor service leads to the perpetuation of low tariffs and insufficient cost recovery which, in turn, leads to under-investment, a deteriorating infrastructure, and poor service. . . . This is reinforced by organizational deficiencies, inefficient management and inadequate skills.

The public undertaking functions as an administration and not as a firm with financial objectives. It is governed through bureaucratic procedures without any kind of incentives: operational agents are not given the means to maintain the infrastructure properly, although they are well aware of the problems; investment decisions are unproductive; technology is obsolete; field data is not correctly or fully reported and the information transmitted to the hierarchy is not reliable (Bijlani 1993). Thus, although rates are very low, this does not fully account for the poor financial and technical condition of the undertaking and mismanagement accounts for a significant part of the problem. For instance, distribution losses between treatment plants and customers are as high as 30 percent to 40 percent and almost half of total consumption is not metered. Finally, water unaccounted-for is estimated to be as high as 60 percent (Rohilla and Datta 1999). Thus, leaks detection and network rehabilitation should be a major priority, which they are not in practice.

This is coupled with the problem of political interference leading to inappropriate decisions that focus on construction issues and more water treatment plants,[11] dams, etc., rather than seeking to improve the existing infrastructure and ensuring raw water availability. Moreover, decisions do not address one of the most salient features of the city, i.e. its fragmented character, which is the consequence of erratic planning. In other words, the decision process is short-sighted and unpredictable.

This rapid analysis of the drinking water supply system in Delhi underlines the complexity of the institutional environment and reveals many failures, in particular, organizational ones. This sector is characterized by the absence of effective regulations, controls or coordination between the agencies concerned. The governance structure has a clear impact on the institutional framework and this partly explains why the municipal undertaking does not succeed in meeting the basic needs of the population.

Individual and Collective Responses to Inadequate and Unreliable Supply

Turning now to the third level of our analysis, we will focus on individuals' responses to the unreliability of water supply. Individuals do play an important role in the architecture of the water system. Through their own arrangements, what we call "decentralized governance structures," they provide alternative modes of supply. However, the social, economic and environmental sustainability of such individual strategies is questionable.

So far, we have assumed that the network supplies the whole city. In fact, the distribution system is discriminatory, in the sense that many areas are not served. Such areas include peripheral neighborhoods (both rural and newly constructed dwellings) and many slums settlements.[12] This is due to discontinuous spatial development and will probably persist along with the growth in the urban population. Both poor and well-off people are therefore affected by the lack of infrastructure or by inadequate supply, but of course not in the same proportion to their respective revenues.

In this context of highly inefficient public supply, people have developed compensatory strategies and alternative modalities of supply have emerged. They can be divided into two categories: formal and informal strategies.

Formal strategies consist of relying on private operators which sell water in large quantities via water tankers (containing around 12,000 liters).[13] Many people also buy bottled water and water in jars, however such strategies are affordable only to high-income households. The major problem with these sources is that water quality is not guaranteed, and some opportunistic firms simply resell public water or sell untreated groundwater. The absence of any regulation in this sector has enabled the emergence of small companies with a short-term strategy. Such companies have taken advantage of a booming market without investing in quality equipment and operate at a low cost of production. On the other hand, companies that set up sophisticated production lines with a view to establishing themselves in the market on a long-term basis have complained of this unfair competition. They were also dissatisfied with the high taxes imposed by the State government in Delhi (bottled water is considered a luxury item) and favored stricter regulations, which, as of 2002, have not been approved. So far, these private ventures, which are a direct result of the inefficiency of the public sector, have not been able to come up with innovative solutions to provide services at affordable prices and to guaranty the safety of water. The solutions they offer are only peripheral and temporary ones.

Informal strategies are strategies which are external to any market structure. Such strategies are developed by poor and well-off households alike. Most of the time, the poorest people still rely on public water via illegal connections onto which they install cheap devices to pump water from the network. This behavior can be described as "free-rider" behavior.[14] Higher income households adopt more expensive strategies: some install electric pumps in order to pump more water from the network thanks to better pressure; some store water in rooftop tanks; some dig tube-wells and rely on groundwater.

Toward a Sustainable Water Supply Service

In this last section, we wish to discuss the implications of our analysis on the key issue of the sustainability of water services (see also Barraqué in this volume).

The Unsustainability of Current Arrangements

All compensatory strategies generate direct investment costs (storage facilities, motors, filters, etc.). In Delhi, the total expense incurred by households for such strategies is 6.5 times higher than what they pay directly to the public undertaking. The aggregate cost of water unreliability at the city level is equivalent to almost twice the amount of the annual expenditure incurred by the former Delhi Water Supply and Sewerage Disposal Undertaking (Zérah 2000).

However, these private arrangements (formal and informal) also generate indirect costs for society as a whole as they contribute to the deterioration in the existing infrastructure through unauthorized water connections. During breakdowns, contaminated water enters the network and exacerbates the risk of waterborne diseases. Regarding groundwater, multiple unregistered private tube-wells deplete the water table. Finally, private arrangements aggravate the water shortage and congestion phenomena. In other words, a system of negative externalities becomes self-sustaining with a harmful impact on the environment and on users' health.

From an economic viewpoint, these decentralized strategies for dealing with the inadequate service are not the most efficient in view of the additional costs that they generate, and they are clearly not sustainable. However, storage solutions, rain-water harvesting and water supply via tankers may offer acceptable temporary solutions provided that a well-defined regulatory framework is implemented and enforced.

Community participation in the management of decentralized infrastructures could also be promoted. Our work and other research suggest that the institutionalization of community participation mechanisms is desirable for at least three reasons. First, this would allow the additional costs of compensatory strategies to be internalized and enable a more equitable redistribution system to be set up. Second, householders would be provided with an effective means for ensuring that the infrastructure is properly maintained. Third, water resources would be more effectively managed, thanks to a demand-oriented approach and by facilitating leak detection. Thus, access rights to water would be secured. However, this would require major institutional changes and, in particular, the democratic representation of all interests, the setting up of agreed-upon negotiation procedures and the abandoning of patronage relationships (Haider 1997; Llorente 2002).

Current strategies are a response to an inefficient service administered by an incomplete institutional environment that is unable to provide suitable incentives. They are affected by the absence of formal rules and this results in a chaotic allocation of the resource. Although they are not sustainable, the existence of such arrangements suggests that reform of the sector should be analyzed in a systemic way and that consideration should be given to the opportunities offered by decentralized governance structures. By a systemic approach, we mean analyzing all interaction between the agents, the resource and the institutional environment. In the case of water, this analysis reveals huge differences between developed and developing countries that preclude the mere transposition of a contractual model without any other kind of consideration (see Figure 10.3).

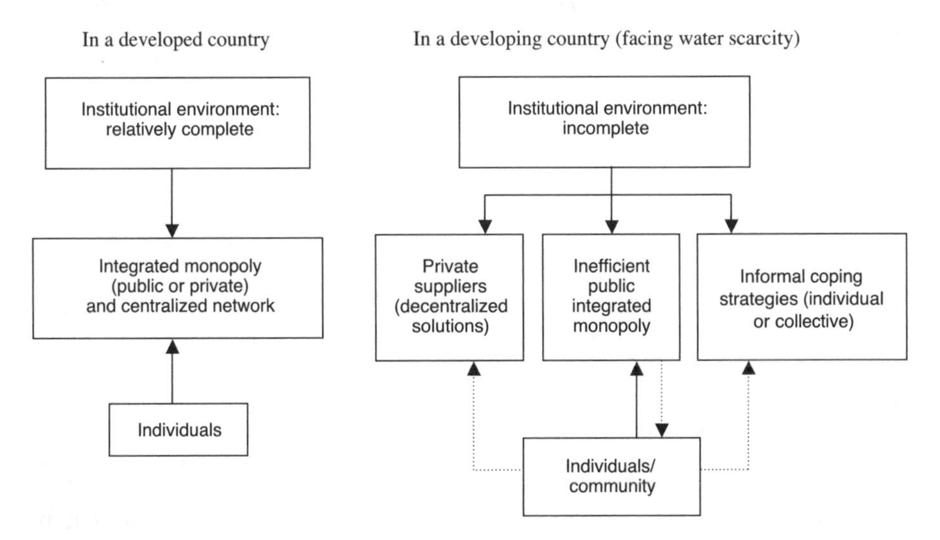

Figure 10.3 *Water Supply and its Institutional Environment: Two Typical Configurations*

Sustainability through the Centralization of Supply?

The situation prevailing in a city like Delhi raises the question of whether a centralized network would be a sustainable solution for all. This model, which draws its inspiration from western industrialized countries, requires a homogeneous city with rational planning. At present, housing and spatial dynamics are chaotic and service provision is thus erratic. The current approach is supply-oriented rather than being based on the needs to be met. It would be more appropriate to take a broader, demand-oriented approach and consider water supply in terms of the service available in the various areas, namely centralized and/or decentralized supply via public and/or private providers.

However, promoting private-sector participation (PSP) in the current context in the form of a single, all-Delhi franchise contract would be useless. It might address some deficiencies, in particular production-oriented ones, thanks to improved management and financial streamlining, however, the operator would have to deal with the fragmented nature of the city, the question of ability to pay in poor areas and the risk of the local authorities not appropriating their profits. Moreover, there is a risk that a private firm would behave opportunistically by concentrating its efforts on profitable areas and delaying network extensions in low density areas. The subsequent price increases would probably be opposed by the population and weaken Delhi's governmental coalition.

In this unreliable context, PSP would not be viable because more fundamental problems would probably remain unsolved: threats of the availability of raw water in the medium to long term, the required upgrading of the infrastructures, interstate water disputes resulting from an inadequate legal framework, the persistence of patronage procedures and the politicized nature of the water supply, etc. In such a context, what is the point of issuing a tender for the construction of a new water treatment plant? In other words, a franchise contract would not by itself provide an effective solution to the broader *public service crisis* that affects water supply in Delhi.

Towards Effective Reform: A Neo-institutional Perspective

Through a systemic approach, we have argued that the roots of the urban water crisis in Delhi lie more in the waste of resources than in a lack of such resources. The depletion of the resource is accelerated by the overexploitation of underground water, massive leakage and inefficient uses. The financial difficulties faced by the public undertaking result primarily from inefficient management and under-pricing. This implies that effective water supply and sanitation policy reform should be comprehensive, and not limited to price increases or a franchise contract. A proper reform should tackle both the inefficiency of the governance structures and the incompleteness of the institutional environment.

Regarding the reform of the public undertaking, two main solutions exist: (1) public corporatization,[15] which could help in transforming current practices into more efficient ones through incentive mechanisms, but which would require a learning phase and time to adapt; (2) franchising, i.e. a public–private partnership with various options for risk-sharing and contract duration in which the local authority retains ownership of the assets. Both options have their pros and cons (Llorente 2002) and, although the purpose of this chapter is not to discuss their comparative efficiency, two comments should be made. First, the results of such a reform depend primarily on the quality of the institutional framework. Second, with either option, three transactional issues remain: the limitation of informational asymmetries, the design of suitable incentives in order to restore accountability and the warranty of reliable commitments to limit interference and the charging of excessive prices. In other words, the key issue is devising contract enforcement mechanisms.

Regarding decentralized solutions, water delivery through tankers constitutes a temporary solution. However, it reinforces economic and spatial inequalities as the price of this service excludes many potential beneficiaries. Institutionalized relationships between some of these providers and public authorities could be established, however, they would require careful consideration of distribution patterns, quality control and pricing.

Community-based programs in low-income areas certainly offer the best opportunities to improve access to water in these areas. To succeed, they need to be designed from the bottom-up, i.e. in a way that effectively involves the beneficiaries of such projects. Haider (1997) argues that several programs have failed because they were designed at the top level in an authoritarian manner, disconnected from the realities on the ground – a classic syndrome of top-down, ineffective planning. Decentralized, grassroots planning appears essential in order to properly address implementation issues. The beneficiaries of water services should be consulted and given fair negotiating powers through their representatives. These arrangements should be part of a comprehensive water management strategy aiming to progressively improve the level of service.

Regarding the institutional framework, the task is not any easier, however, several measures could improve the existing framework. The first priority is to simplify the architecture by redefining the allocation of responsibilities both vertically (e.g. between the various administrative levels) and horizontally (e.g. between water-related issues). For example, appropriate functional relations in agency assignments could be established between: (a) water and land use, so as to conform to planning objectives; (b) surface and groundwater, for conservation purposes; and (c) water quality and quantity, in order to promote preventive measures rather than curative ones, etc. In this view, central nodal agencies still have an important role to play in designing long-term policies and planning, however, such policies would be better enforced than they are today.

Policies and enforcement of regulations require tackling problems such as corruption and patronage, etc. How can behavior be changed? How can a sense of probity be introduced into administration and politics? This moral question goes beyond the scope of this chapter, but the answer probably lies, partly, in a more democratic decision-making process. The 74th amendment to the Indian constitution does call for local representative structures, however, their implementation is constantly being postponed. The lack of political will is often identified as the main obstacle to reform.

In conclusion, India, today, is in a learning phase regarding regulation issues. Nevertheless, it is clear that studying water reform solely from the standpoint of privatization is not suited to the Indian context. The first priority is to focus on the reform of public action and to redefine the role of the different institutions involved in the governance of the water system. The key question is thus to find incentives that would lead the government and public agencies to perform these new roles and become accountable. This is undoubtedly the main issue underlying the largely misleading public v. private debate.

Notes

1 According to the World Bank (1998), the Indian urban population is growing at an annual rate of 3.1 percent which is significantly higher than the overall population growth rate (2 percent). It is estimated to reach 658 million by the year 2025.
2 We observe a considerable degree of stability with regard to the forms of industrial organization in developed countries, with public arrangements being prominent. Only a few Western European countries manage their water services through private delegation contracts. The UK is well-known for its fully privatized sector, while approximately 85 percent of US water supply firms are publicly operated (Nelson 1997).
3 The first paragraph of the National Water Policy Act of India (2002) states that: "Water is a prime natural resource, a basic human need and a precious national asset. Planning, development and management of water resources need to be governed by national perspectives."
4 This corresponds to what Williamson (1985) calls "site specificity," which he views, in the case of network industries, as the most important asset specificity, the predominant factor of integration of interdependent activities within a single firm.
5 Externalities occur when the actions of one party or agent affect the welfare of other parties or agents in a way that is not mediated through markets. Externalities are considered market failures that require coordination between agents in order to redress them.
6 A municipal corporation is a fully representative body to which councilors are elected every 4 years.
7 Estimates of leakage range from 30 to 40 percent.
8 The following data is taken from a survey conducted in four Delhi slums (sample: 110 households) between April and June 1998 (see Zérah and Llorente 1999).
9 The National Capital Region includes the Delhi Union Territory (1,483 km^2) and a part of the neighboring states (Haryana, 13,413 km^2; Uttar Pradesh, 10,853 km^2; Rajasthan, 4,493 km^2), i.e. a total of 30,482 km^2.
10 The following observations are taken from a field survey carried out in 1999 and are corroborated by other studies (World Bank 1998; Mehta 1999) and by similar analyses in the electricity sector (see, for instance, Ruet 2001).
11 A new water treatment plant with a capacity of 635,000 m^3 per day is under construction in East Delhi (Degrémont, a subsidiary of Lyonnaise des Eaux France, recently won the construction and ten-year operating contract).
12 Some of these have legal status while others result from illegal land occupation (squatter settlements).
13 Note that the public undertaking also provides free water in some settlements using public water tankers, however, these only deliver rarely and in small quantities.
14 A situation in which several different parties can use a resource for their individual benefit without paying for it because property rights are not sufficiently well-defined or enforced to ensure that individuals bear the full costs of their actions.
15 Corporatization consists in giving a public service department a form of autonomy in order to induce it to operate as if it were a private firm in a competitive, efficiently

regulated market. In an Indian context, this notion mainly entails cost efficiency and the absence of external discretionary power.

References

Bijlani, H.U. (1993) *Managing water resources to meet megacity needs – Case study of Delhi, India*, final report for the Asian development bank, Manila, Philippines: Asian Development Bank, p. 306.

Central Ground Water Board (CGWB) (1996) *Development and augmentation of groundwater resources in the National Capital Territory of Delhi*, Indian Ministry of Water Resources.

Haider, S. (1993) "The politics of providing basic amenities to the urban poor," in B. Mohanty (ed.) *Urbanization in developing countries*, New Delhi: Institute of Social Sciences and Concept Publishing Company.

—— (1997) "Community participation in basic services and environmental protection: case study of a Jhuggi-Jhompri cluster," *Man & Development* 19, 4: 158–88.

Laffont, J.-J. and Tirole, J. (1993) *A theory of incentives in procurement and regulation*, Cambridge, MA: MIT Press.

Llorente, M. (2002) *Une approche néo-institutionnelle de la gestion urbaine de l'eau à Delhi: quelle régulation pour quel service?*, PhD dissertation, Nanterre (Fr.): University of Paris X, p. 344.

Mehta, M. (1999) *A review of public–private partnerships in water and environmental sanitation in India*, New Delhi: Department for International Development.

Milbert, I. (1998) *Politiques urbaines à Delhi*, IUED report, Genève (CH): IUED (Institut Universitaire d'études du Développement), p. 61.

Ministry of Urban Development, India (1988) *Report of the National Commission on Urbanization*, Vols I–II and Interim Report, New Delhi: Ministry of Urban Development.

Narang, S. (1997) "Agenda for people's participation," *Down to Earth* (30 June): 56–7.

Nelson, Michael (1997) "Municipal Government Approaches to Service Delivery: An Analysis from a Transactions Cost Perspective," *Economic Inquiry* 35: 82–96.

North, D. (1991) "Institutions," *Journal of Economic Perspectives* 5: 97–112.

Rohilla, S.K. and Datta, P.S. (1999) *Delhi's water and solid waste management emerging scenario*, New Delhi: Vigyan Prasar, p. 76.

Ruet, J. (2001) *Winners and losers of the state electricity board reforms: an organisational analysis*, CSH Occasional Paper 1, New Delhi: Centre de Sciences Humaines, p. 85.

Sinha, S.D. (1995) "Environmental restoration for providing clean water: Delhi as a test case," *Spatio-economic Development Record* 2, 6, November–December: 8–18.

Suresh, V. (1998) "Indian experience in urban water supply and sanitation," paper presented at ESCAP sub-regional workshop on private sector involvement in the water supply and sanitation, New Delhi, April.

Williamson, O.E. (1985) *The economic institutions of capitalism: firm, market and relational contracting*, New York: Free Press.

—— (1993) "Transaction cost and organization theory," *Industrial and Corporate Change* 2: 107–56.

World Bank (1998) *India water resources management sector review: urban water supply and sanitation report*, report No. 18321 to the Rural Development Unit, Washington, DC: The World Bank.

Zérah, M.-H. (2000) *Water: unreliable supply in Delhi*, New Delhi: Manohar Publishers, p. 168.

Zérah, M.-H. and Llorente, M. (1999) *Institutional innovations in the Indian water supply sector*, final report to the Centre de sciences humaines, New Delhi: Centre de Sciences Humaines.

CHAPTER ELEVEN

Not Too Much But Not Too Little: The Sustainability of Urban Water Services in New York, Paris and New Delhi

Bernard Barraqué

When Europeans spend a few days in New Delhi or Mumbai, they do not necessarily bring along chlorine pills to purify the tap water. It would taste too awful. They just drink hot tea, sodas or mineral water. Now, an increasing number of people in Europe and the US are willing to pay 200 to 500 times more for bottled mineral or purified water than they would pay for tap water, and to carry it home, so as to avoid catching diseases from the public water supply (PWS). Moreover, now that chlorine has been found to be responsible for secondary carcinogenic effects, its good old taste does not make US water drinkers feel secure any more! Taken together, these three pieces of information raise the issue of the sustainability of water services: can the developed world ultimately afford a public drinking water supply plus the related sewage collection and treatment costs, or will the deficient and substandard water services of large third world cities become the norm, together with bottled spring water or purified water?

In California, 30 years of discussions concerning environmentally friendly growth, together with privatization and the end of government subsidies, have led electricity supply or telecommunications systems to the verge of disintegration (see Rochlin in this volume) and water systems to a point where it has become impossible to postpone the next water shortage through further expansion towards northern water resources. In Europe, the increase in environmental directives (laws) and liberalization reforms in public services (based on so-called "full cost recovery") have simultaneously led to more instances of non-compliance with drinking water standards and to larger water bills, while for other reasons, and for the first time since the inception of water services in the nineteenth century, water volumes sold are stagnating and even going down (for France and Switzerland, see Barbier 2000). This unprecedented reduction in demand requires unit price increases to cover the fixed costs. The ultimate result is the growing distrust of customers in their water utilities. In developing countries, the situation is clearly appalling: contrary to the expectations of the "water decade" that started in the mid-1980s, connection levels did not improve much, and the World Water Council has calculated that in order to achieve the connection of urbanites to both PWS and public sewage collection and treatment (PSCT), our planet was short of *a mere* $100 billion per year over a 25-year period. Now isn't it time to reflect on how we got into this incredible mess and where we are going to go from here?

Geographic and Historical Differences

In terms of drinking water uses, there are also three different worlds, however, the geographical breakdown is different from the breakdown from a development perspective. At one end are the large industrial countries which have developed on the scale of continents and where large hydraulic projects bring huge amounts of water for use by metropolises and for other uses, in such a way that allocation conflicts could be staved off: domestic consumption is more than 500 liters per capita per day (lcd), or 130 gallons per capita per day. In Las Vegas, demand for potable water goes up to 1,200 lcd and requires an ever greater share of Colorado water for more and more lawns and swimming pools in the middle of a desert. At the other end of the scale, there are the fast growing cities in developing countries, where the rich have the same consumption patterns as in the US, but where the poorer part of the population is only half connected – or not connected at all – to "inconstant" supplies, i.e. supplies with frequent service interruptions which are not compliant with drinking water standards (Zérah 1997). Average consumption is under 80 lcd; sometimes it does not even reach the 40 lcd that the 1992 Rio conference on sustainable development decided was the minimum amount that should be delivered regardless of the capacity to pay. In between, there is Europe, where the relative scarcity of resources, demographic congestion and early policies of universal service have led to a moderate, yet comfortable level of consumption at 150 ± 40 lcd. However, water prices here are usually three times higher than in the US or in Canada.

In order to explain these three broad types of situations, one has to analyze not only geographical, but also technical and financial historical considerations. In the nineteenth century, or rather until Koch and Pasteur's discoveries had been popularized, PWS developed on the assumption that water should be drawn from natural environments far from cities. Large cities, in particular, would have to get water from further and further away. This was made possible by the ability of municipalities to obtain "cheap money," especially from the early popular savings banks that they controlled. Their bonds were popular with the public. And on top of this, governments were willing to subsidize projects. Take the example of Glasgow:

> direct municipal provision seemed to offer several advantages to the city. The existing private company had . . . an outdated infrastructure [and] consequently was unable to cope with the demands of the rapidly growing population . . . Moreover, the company was not in a position to raise the necessary capital for improvements, unlike the Town Council, whose extensive community assets made it eminently creditworthy. Public accountability meant that unpredictable market forces could be over-ridden, and a stable service provided . . . Loch Katrine was located in the Perthshire highlands, some 55 km from Glasgow, and thus well away from the polluted city . . . The official opening by Queen Victoria on an appropriately wet autumn day in 1859 was an event of enormous significance for Glasgow . . . Loch Katrine was unquestionably the prime municipal showpiece for the city, combining the wonders of Victorian technology with the nurturing quality of pure Highland water.
>
> (Maver 2000)

Joel Tarr (1996) has illustrated this broad approach in the US: getting cleaner water from further afield on the one hand, and using rivers as sewers on the other, this latter decision relying on the assumption of natural dilution and self-purification of rivers.

In order to complete the connection of the whole population to PWS, many cities not only took over the utilities created by private companies, they also changed the financing system completely. Indeed, creating a local water tax or making the connection to the PWS compulsory and basing water rates on property values provided the money that was needed and generated massive cross-subsidies from richer to poorer urbanites. The example of Montreal illustrates this (Fougères 2002).

This approach remained the dominant one in the New World, and was also extended to the rest of the world after the Second World War, due to the twin circumstances of international financing institutions offering cheap money and the existence of various (Keynesian or socialist) forms of support for government intervention in infrastructure provision. In the 1950s and 1960s large hydraulic projects were increasingly devoted not to cities, but to irrigating agriculture for the export market. Even today, many states in developing countries still base their water policy on large water transfers, so as to indirectly subsidize the production of irrigated cash crops in order to become integrated on the world market. In some cases the ceiling of extractable water resources has been reached, and the present crisis offers the possibility of confirming the non-sustainability of these past policies.

The Response to a Technical Reverse Salient[1]: From Further Quantities to Nearer Qualities

However, a similar crisis occurred a long time ago in (Northern) Europe: growing population densities and smaller natural resources increased competition for pure water resources, while the development of bio-chemical analyses revealed the growing extent of contamination. As irrigation is unnecessary in Northern Europe, there was not so much a resource quantity problem as a quality problem. In the end, it was decided that, whatever source it came from, water should be filtered (end of nineteenth century), and later chlorinated, ozonized or disinfected through GAC (granulated activated carbon) beds (around the First World War). But then, if water was to be treated, taking it from the river just upstream from the cities would not change much in terms of public health and would save a lot of investment. So in that still early period, large European cities changed their strategy from one of investments aimed at increasing available quantities, to investments aimed at improving quality. This, of course, resulted in a significant rise in operating costs. However, the status of the domestic delivery of pressure water changed from that of a luxury good to a normal commodity and made it possible for customers to pay water bills to cover these costs.

This is exactly what happened in the city of Paris a century ago. From the time of Napoleon III, the idea had existed that Paris should get water from distant sources. Indeed, the work of the engineer Belgrand, under Baron Haussmann's general dedication to public services, was turned towards securing longer distance sources of water (around 100 km away from Paris). It was even felt that one day the capital city would have to get water from the River Loire, even though this river has very low flows in the summer, precisely when water demand is highest.

Then, in 1890, an engineer named Duvillard came up with a project to draw water from Lake Geneva, 440 km away from Paris! It sounded like a far-fetched project, on account of the Alps and the international character of the River Rhone – would the Swiss agree? Yet, in fact, it was technically quite simple, even at that time. Proponents of the project soon came up with all sorts of arguments to convince the Paris city council and the State: a "capital of the world" would need at least 1,000 lcd to have more luxurious fountains, more street cleaning[2] and better domestic comfort and hygiene. Besides, such a quantity of water would extend navigation possibilities in drought periods, help flush waste water from the new sewer system away to the River Seine and then to the sea, and make other water resources available for local economic development. In the end, they argued, such an enormous water transfer would make the Paris PWS reliable forever, and the bigger the transfer, the cheaper each cubic meter.[3]

But while proponents were finalizing the studies, an epidemic broke out, and it was found that one of the distant natural intake points (the Loing springs) was to blame: thus, even distant, "pure" water could be contaminated and should be filtered and treated. In 1902 Paul Brousse, one of the founding fathers of so-called French "municipal socialism" (equivalent to what was derided in England as "water-and-gas socialism"), inaugurated the new water filtration plant at Ivry, just upstream from Paris (the same one which was redesigned a few years ago to serve as a showcase for French water technological know-how): a long-term choice was being made – it was decided to chlorinate water after the First World War. Water demand was growing incrementally at this time and the "big jump" then appeared too risky. Finally, the Lake Geneva project was ruled out by the Paris city council for national defense reasons: what would happen if during the next war the Germans invaded and cut off the aqueduct? Even though this decision did not help much in the next war (are there ever real water wars?), it is clear that quality investments replaced quantity ones, just as chemical engineering replaced civil engineering.

After the Second World War, the prefect of the Seine *département* (district) took advantage of a severe flood to obtain an agreement to construct three large upstream reservoirs on the Seine, the Marne and the Aube rivers in order to increase summer water flows and meet Paris's water demands, even during very serious droughts (such as that of 1976). Interestingly enough, a fourth upstream reservoir was planned by Mayor Chirac's councilors in the 1990s, but it was abandoned because the giant water supply companies maintained that they would have had to purify the water anyway and because water demand in Paris went down by 13 percent between 1991 and 1997 (Cambon-Grau 1999).

Likewise, the invention of water treatment (chlorination was developed during the First World War) allowed many cities to use nearby surface water, and thus to complete the networks and serve the population using mostly local resources. This is one of the good reasons for generalizing municipal control over utilities, at least in temperate climates. Ultimately, the second best solution would appear to be an inter-municipal joint board. The decision to treat water also meant growing operations costs and helped promote the idea that water was a commodity to be paid for by water bills, thus increasing the financial autonomy of the utilities. This commodification was more readily accepted during the twentieth century when urbanites grew accustomed to having tap water. However, the initial infrastructure, which had frequently been subsidized, would eventually need to be replaced, confronting utilities with a major financial issue.

The Crisis of Municipal Water Supply Services: A Financial Reverse Salient?

In France, as in other Northern European countries, important efforts were made regarding city sewage collection from the 1950s onwards, and on sewage works from the 1970s onwards. To make investment easier, it was decided to change the status of PSCT from a public health system, paid for by local taxes, to a public service similar to PWS, paid for by water bills. But over the same period,

PWS itself became a mature business, i.e. it had to face the issue of renewing aging infrastructure without the benefit of subsidies. This is the fundamental reason why municipalism evolved in various ways towards entities with a private legal status: traditional public accounting could not depreciate the assets nor make renewal provisions while private accounting could. Thus, PWS, and later PSCT, slowly turned towards depreciation and provision practices, and this of course meant a rise in water bills. Besides, governments are now under the influence of economists who argue in favor of full, or at least fair cost pricing[4] and subsidies have been phased out. In turn, water bills have risen dramatically (in Paris and the rest of France, bills nearly doubled in real terms between 1990 and 2000) and an increasing number of large users (industry, services) either change their processes or try to reduce wastage. This explains the recent stagnation in volumes sold. In some countries, even domestic consumers have reduced their demand for PWS through changes in fixtures and domestic equipment, different garden designs and rainfall storage or other alternative sources of water for non-drinking uses.

At the same time, water suppliers have discovered that it is going to be increasingly hard to produce water that complies with drinking water standards all the time at a reasonable cost. The control of eco-toxicologists over standard production tends to favor a traditional "no-risk" strategy (Lave 1981) without taking the implied costs into account: in Europe, lowering the lead content from 50 to 10 μg/l will cost up to $35 billion, while there is no evidence that the previous standard was responsible for lead poisoning. The increasing number of criteria[5] is gradually making the situation overly complex. Byproducts of chlorination cause a very small risk of cancer and there are many other examples. Year after year, the media report a growing proportion of people receiving substandard water, even though the treatment is improving in the long run (for a history of drinking water criteria and the resulting outcome today, see Okun 1996). To lower the risk of being unable to make water meet the requisite standards, along with local, national and European authorities, water suppliers have begun to use a new strategy: land-use control in catchment areas where groundwater is extracted by PWS; this often implies changing over to biological, or at least non-nitrate, non-pesticide agriculture and compensation programs for farmers (Brouwer *et al.* 2003). This policy turns out to be cheaper than the sophistication of water treatment and allows for a positive-sum (win-win) game with farmers. Yet, this seems a long way from a new and more sustainable balance. And worse, in the meantime, drinking water criteria are being reinforced regularly as new risks are discovered and the spiral of treatment costs and their negative effects continue unabated.

In any case, contrary to what some "hydro-schizoid" water engineers seem to think in Spain (Llamas 2001), the present time is not appropriate for large-scale, long-distance water transfers (Barraqué 2000). What happened in California, beginning in the 1970s, has now become general practice in Europe: it is becoming increasingly more difficult to build dams because environmental movements have been joined by economists and liberals who advocate full-cost payment of water infrastructures by their beneficiaries. The new rationale now seems to be: "save first and then manage the demand, there is no cheap money in sight for subsidized water transfers." Copenhagen is not going to buy water from Sweden, Puglia is not going to have Albanian water, London will have to reduce wastage drastically before it can purchase water from Scotland. And many other fancy projects on other continents are dying like "hydro-dinosaurs".[6]

New York City, like many US and Canadian cities, could follow a different path from European cities because of the abundance of clean water. The tradition grew up of using a lot of water and of taking it from further and further afield, while protecting water intake points through extensive land-use control. Yet, US cities may have to catch up with the European side of the story, as clean natural resources are not immune from cryptosporidium and other new (lethal) diseases. A US Environmental Protection Agency panel of experts concluded that New York should not be accorded any further derogation in relation to the need to filter and treat water extensively, while city engineers were arguing that increased land use control would suffice (Okun *et al.* 1997; Ashendorff *et al.* 1997). The new treatments requested are likely to seriously increase water prices,[7] which might lead to a collapse in water demand and to a further question: in that case, why not just pump the water from the Hudson River, and forget about the damned *Quebecois*?

Hence, in Europe, because of population density and environmental issues, the water industry has gone through three "stages": it has had to progressively supplement the supply, i.e. quantity approach (hydraulics and civil engineering; first stage), with a quality approach (water treatment and sanitary engineering; second stage) and, more recently, with resource protection and demand management (environmental engineering; third stage).

Eurowater and the Three Es of Water Services Sustainability

But can the Europeans afford their water policy? This was an increasingly important issue in the 1990s with the first estimates of the cumulative investments implied by the anti-eutrification directives adopted in 1991 (directives on urban waste water and nitrates from agriculture) on top of the existing ones. Water policy has been increasingly criticized by the public, the press and politicians. Paradoxically, criticism has been much stronger in Northern European Member States where a complete infrastructure exists, than in the Southern States, where equipment is still being placed in service and where water prices are still a long way from matching costs (which, in turn, means government subsidies or infrastructure degradation). It shows that *rationalization* (both in economic and environmental terms) cannot really start until *rationing* is over.

Furthermore, the issue has also been developed between some Member States and the European Commission. It was felt that supplementing traditional standards based on emissions control (pollution discharges, drinking water criteria) with another type of regulation based on the desired ecological quality of the aquatic environment would be over-costly and, besides, all these regulations would be impossible to meet at the same time, with the result that governments could be sued at any time by their own citizens for non compliance with the legislation. This resulted in a postponement in the issuance of a projected directive on ecological river quality until a framework directive had been drafted and discussed, so as to introduce coherence and cost-savings. The framework directive (EEC 2000/60) was issued in October 2000, and some of us wonder if the European Commission, Parliament and Council are not setting an impossible target, i.e. "cleaning all water in fifteen years," as the US Congress did in 1972. On top of this, the European Commission and Council officially support full-cost pricing, without providing any convincing proof of the mere practicability of such a strategy.

However, the issue of costs may have been biased: the costs of implementing regulations have generally been calculated on the basis of technologies that are now well mastered, but which were developed by sanitary engineers before the emergence of environmental issues and under a different rationale vis-à-vis risks and costs. The high costs incurred today might well be due to the inability to develop environmentally innovative technologies and, above all, to develop strategies other than "supply side" or "end of pipe" ones, i.e. demand-side management, land use control and integrated planning (see Moss 2000). Besides, Southern Member States, with a different climate and different population patterns, may need to develop specific technologies rather than copying those of Northern States, even though they receive support for these via the European Union structural funds. However, it will obviously be difficult to get some of these Southern Member States to overcome this new technological reverse salient before some severe crisis shows that they have reached it. These issues were addressed in comparative research on the forecast costs for complying with the most costly directive; that relating to urban wastewater collection and treatment (UWWD), issued in 1991.[8] The Eurowater research project gave rise to an analysis of water engineers' attitudes vis-à-vis environmental policy. More broadly, we have evidence that it is not engineers alone who are to blame, but the kind of traditional relationship they maintained with elected authorities and the public: who would want a "substandard" septic tank in their gardens when everybody tells them that PSCT is the only real solution (Barraqué 1998)?

In a follow-up to this research for the European Union, the same partnership of European water policy analysts has addressed the issue of future water services (i.e. PWS and PSCT) in light of the UN's definition of sustainability: how do we reconcile economic, environmental and ethical/equity sustainability criteria, i.e. the "three Es approach?":

- Economically: how is the enormous capital accumulated in water services technologies maintained and renewed in the long run? If we do invest enough, what is the impact on water bills?

- Environmentally: which extra investments are needed for the sake of public health and the environment, and, when the cumulative costs are too high, are there alternatives to the traditional systems?

- Ethics and equity: if all "sustainability costs" (long-term maintenance of the infrastructure, environmental protection and user costs) are passed on to consumers, can they afford this, and is it politically acceptable?

The first two issues are, in fact, inter-related because of the limited capacity of governments and cities to deal with unsubsidized investment: investments to replace ageing infrastructure and new environmental investments do overlap, however, both have to be made without subsidies. If England and Wales were to rebuild their water services infrastructure completely, they would have to spend £189 billion! Some of these assets can be depreciated over 100 years, but others only over 30, or even over 10 years. With the present centralized regulation system, OFWAT, the UK water regulator, can calculate what investments should be made and when, and which price should result. Other countries are also trying to do this despite the decentralized organization and small size of undertakings. But many experts think that we are simply in the process of "consuming" the initial capital, in particular for sewers. Besides, water prices in the UK are reasonable compared to those in other countries. However, to make privatization attractive in 1989, the government cancelled the debt of the previous regional water authorities and, on top of that, offered a so-called "green dowry" to help the new companies face the "nasty European directives." Altogether, what everybody else, apart from Thatcherites, would call subsidies amounted to £6.4 billion, i.e. more than the French or German governments gave away to their water services over the last 20 years! In her thesis, Bakker (2004) shows that British privatization put the water industry into a fundamentally unsustainable situation, which will be increasingly obvious in the coming years, more than 20 years after it occurred. This is why the issue of cost recovery, which was raised by the new water framework directive is so unclear: if you want to analyze economic sustainability seriously, you have to use a 30- or 50-year perspective to see how subsidies during that period influence today's price. If subsidies are removed now, the effects on water prices will not be felt for at least 20 years.

We have also found that water suppliers or water authorities, under various modes according to the politico-historical culture, have organized averaging out or cross-subsidy mechanisms to limit the impact on water prices of heavy and lump-sum, but long-term investments. Some do it through spatial integration,[9] others through temporal averaging (earmarked funds, water banks, modernization of public accounting to allow for depreciation, etc.). The Germans also do it through an original institution, the *Stadtwerk*, i.e. a single, formally private but publicly owned company that runs several technical networks in a given city: this model was criticized by the World Bank, but on the basis of incorrect data (Barraqué 1998). And, last, social forms of averaging are well known, even if they are not always presented as such: for instance, paying for waste water via local taxes makes this service more expensive for those wealthier people with a large house. Under municipalism, water suppliers did not pay much attention to the detailed breakdown of potable water uses and to the distributive effects of tariffs, since they wanted the best quality in unlimited quantities to serve all purposes (in a *commonwealth* vision); those in charge of modern water supplies are still reluctant to really study distributive effects: they want water services to be commodified (through billing instead of taxing) for financial reasons, and they simply assert that it is more equitable that way. However, the English and French examples show that commodification is dangerous for water companies, as consumers will not readily accept that a good service is costly. They will be even more reluctant to accept the fact that, if they save water, their unit price (per cubic meter) will probably go up! Thus, the opening up of the traditionally closed PWS policy community to a whole range of newcomers, in particular the general public, makes water engineers feel awkward and insecure. But is there any alternative to this opening up if what is at stake is the public's general confidence in the service?

A Social Reverse Salient?

It is therefore the third issue, the third E of Eurowater's sustainability, the ethics/equity dimension, that is the most crucial today: is social and political acceptability still possible in the long run, or are we heading towards a collapse of the networked model for water? Until water suppliers got involved in demand management and pricing closer to real costs, there was a broad acceptability of water services. However, this was based on what Martin Melosi calls "out of sight, out of mind." Now the limitations of the supply-sided model forces managers to study their demand like any normal business, i.e. through a market-based approach. The breakdown of demand for drinking water and the development of sector-differentiated uses have to be studied. Nevertheless, this may not be sufficient, and utilities may well have to involve the public in their studies. For instance, claiming that metering is more equitable and efficient (because saving water can thus be rewarded) may prove to be difficult to explain to residents of condominiums in central cities, because water use is largely determined by appliances and because there is little elasticity of demand to price (Chesnutt and Mitchell 2000).

Yet, even though local econometric studies show little elasticity of domestic demand in relation to price, there is a slow reduction in use per capita taking place in the developed world. In the US, this may be obtained simply through information policies and subsidies to individual conservation measures (Dickinson 2000). The point is, to make changes gradually so as to allow people to adapt to much higher water prices over sufficient time. In Northern Europe, lower consumption and higher prices reduce the scope for change. Yet, there has been such a large reduction in water use (mostly from large users and for outdoor use) that the only way to maintain the cost recovery principle is to raise unit prices. Due to the lack of information given to users, in particular on the difference between short-term and long-term sustainability, and on the real causes of (formal) privatization, the situation generates a growing distrust of systems that had reached their equilibrium under a "municipalist" style of management.

Southern Europe illustrates this point differently, since the incomplete nature of the infrastructure prevents prices from being raised to cost recovery levels, which generates irrational allocation. Thus, Almeria might get water from the Pyrenees or the River Rhone, at a cost that is ten times what the city would have to pay to buy this water from farmers. Based on a model initially developed under the dictatorship, but now involving European subsidies, Spanish farmers overexploit the aquifers and request increasing volumes of surface water transfers to grow tomatoes and strawberries, part of which will be bought as a surplus by the EU – farms are largely owned by rich Northern European agribusiness. If they were not encouraged by this uneconomical set up, farmers would make more money by not working and by simply reselling "their" water. This is what happened in California, and the only problem was that, in order to be accepted, this arrangement had to be called a "water market."

We have to face up to the fact that the commodification of water services is very dangerous if it is not done within a collective relearning process away from the "out of sight, out of mind" mindset. For instance, there is a trend in Europe towards encouraging the generalization of individual water metering and billing even in flats in small buildings for the sake of more equity and of allowing people to make trade offs between saving and paying, etc. Yet, in most experiments, individual metering does not induce a significant change in consumption patterns, in particular after a few months, and in many cases the savings of the most thrifty are offset by the yearly cost of the meter itself (depreciation, reading and separate billing). Thus, consumers are furious when they see their water bills rise when they had been told they would save money. And distrust grows once more.

One of the most interesting recent cases is the decision of the government of Flanders to implement the Rio Agenda 21 to the letter, and to give away an initial volume of water free of charge. For practical reasons, the free volume was set at 15 cubic meters per capita per year, while extra volumes would be charged in such a way that water suppliers would generate the same income as before. The consequences have been studied by the Social and Economic Council of the Region of Flanders (Van Humbeeck 1998). First, water utilities have seen their total volumes shrink, because increasing block tariffs have made units of water consumed beyond a certain threshold very expensive. The water utilities suspect that people reinvest in cisterns and in private wells for watering gardens etc., but in the meantime, they are obliged to raise unit prices. Second, it turns out that richer families pay a little less than before, while poorer ones pay a little more. This is due to specific socio-demographic conditions: richer families are larger than poorer ones, and extra children do not consume enough water to generate a bigger bill (it should be recalled that the free initial volume is given per person). This ultimately quite complex case illustrates how little we know about domestic water use. If we are going to tackle the present crisis in water services, it will not be simply with an economist's toolbox and solid moral sensibilities. We need anthropologists, sociologists, historians and geographers, etc. The Paris water utility SAGEP has employed an anthropology PhD student to understand the relationship between Parisians and tap water (Euzen 2002). Attitudes are so variable and culture-specific that it would be very difficult to find any economic rationality behind any ostensible price elasticity.

Conclusion

If there is more and more discussion of how to reach a better compromise between the economic, environmental and ethical dimensions of water service sustainability in the EU and in the US, the development of new tools and indicators may give us some confidence for the future. However, what we are missing is an indicator translating the degree of overall confidence of the general public in the systems and those who formulate water policy. But the real threat to water services is illustrated by the situation in Eastern Europe and in large third world cities: when the utilities are not fully reliable, in terms of quantity, continuity and quality (in particular, sanitary quality), the compensation strategies adopted by various groups of users tend to *increase* the uncertainty and the unreliability of the services. In such a case, trying to apply the new economic approaches suited to the mature systems of developed countries might simply be catastrophic. Indeed, it is clear that privatization and even public–private partnerships at state or regional levels will not help to make good water services universal. Perhaps it is simply impossible to improve deficient utilities in developing countries and maybe they are here to stay, together with bottled mineral water. Zérah (1997) referred to a striking case in New Delhi. However, if we look at this example carefully, there is no significant difference between this system and the kind of poor services and private alternatives one finds in Greek islands.[10] And, ultimately, if global confidence in the PWS of developed countries goes on falling, and if water volumes sold decrease significantly, we may end up with increasing irregularity in the service, so that it would be the systems of the large third world cities that would finally prevail in the long run (Barraqué, 2001).

In Durban (RSA), *Lyonnaise des Eaux* has been quite successful with the introduction of special chip cards to buy good quality potable water from fountains: they make some money, users no longer suffer from gastro-enteritis and the company even avoids the traditional tribal control over the wells. In Buenos Aires, the same company currently delivers cheap bulk water at the entrance to the *barrios*, instead of desperately trying to meter each user and fighting for bill recovery (see the discussion on bill recovery in Schneider-Madanes, this volume). However, in both examples, there is something debatable: are poor third world populations bound to make do with these alternative partial services? I would argue that the ultimate reason why they have these alternative types of PWS is that there is no municipality with sufficient legitimacy and capacity to build the necessary reciprocal confidence between the company and its water users.

So let us look to the future of municipalism? In my view, the only way to get out of the *Lyonnaise des Eaux* dilemma is municipal involvement, just like in the good old days before local welfare and public economy had been thrown into the dustbin by liberal economics. What municipalism achieved was to channel the savings of the upper and middle classes into the financing of a long-term system of solidarity for all, based on a public economy of urban services and sometimes involving the participation of the private sector, but without privatization. Indeed, the territorial dimension of the issue is more important than the simple public v. private debate. If we do not want to lose hope for third world cities, we will have to invent similar mechanisms at appropriate territorial levels, depending on national/local citizenship traditions and community cultures. Such a "subsidiary" system would certainly offer better guarantees for national and international public investors which, in turn, would result in access to cheaper money for water systems. This may not fully solve the social reverse salient we have identified in this chapter, but it is certainly part of the solution; in any case, it is more important than international financial institutions think.

Notes

1 In the development of a system, "reverse salients" result from the uneven growth of the system's components. Thus, at a given point in the system's history, a reverse salient is a component that prevents the system's further development. A reverse salient may be technical, organizational, financial, social . . . (see Hughes 1983: 14).

2 Paris is one of the few cities in the world where streets are washed clean: for political and business reasons linked to Haussmann's decision to merge suburban communes and expand Paris from 12 to 20 *arrondissements* (boroughs), it was decided that water for public purposes would be produced by the city and delivered free of charge through a public network, while a second network would serve domestic and other private needs in return for the payment of water bills. This is why Paris still has two PWS systems, one potable and the other non-potable by today's standards. The non-potable network produces barely filtered Seine water to flush the sewers, to supply the lakes in the Boulogne and Vincennes parks and to clean the streets. Other public uses, such as fire hydrants have been abandoned because of unreliability, lower pressure, sprinkler clogging, etc.

3 It is indeed very amusing that the same type of arguments have recently been raised by the advocates of a Franco-Spanish project aiming to transfer water from the Rhone to Barcelona (350 km). This example also shows the indirect impact of irrigation: Mediterranean cities have to get water from ever further afield because all local surface and ground water is given away virtually free of charge to farmers by central government projects.

4 Fair cost pricing means that the principle of cost recovery is accepted but only partly implemented, i.e. bills include only part of the cost of capital depreciation and do not internalize environmental and users' costs.

5 Bacteriological criteria now form only a small part of overall criteria for drinking water standards; heavy metals and micro-organic compounds bring the total to 63 criteria in Europe and 84 in the US.
6 Case studies are available at: www.hydrodinosaurs.fr.st.
7 In particular, if metering is introduced to replace the outdated frontage rates system, the negative redistributive effects could be quite significant, as pointed out by Netzer *et al.* (2001).
8 Eurowater is the name of this research partnership funded by DG XII, LAWA in Germany, the NRA in Britain, and the Gulbenkian foundation in Lisbon. The partners were Tom Zabel and Yvonne Rees of the WRC in England, Jan Wessel and Erik Mostert in the River Basin Administration centre in Delft T.U. (Netherlands), R. Andreas Kraemer and his colleagues at Ecologic, an environmental policy consultancy in Berlin, and my team within the LATTS, a social science institute in the École Nationale des Ponts et Chaussées. The partnership is led by Francisco Nunes Correia, a hydrology and environmental policy professor in Lisbon's civil engineering faculty.
9 And this is what happened in Britain: before the ten regional authorities were centralized in 1974, and later privatized in 1989, there had been a sustained concentration process in progress since the Second World War.
10 In some Greek islands, people rely first on their wells and cisterns, and then turn to the PWS for additional uses. Therefore, the demand for public water is highly variable, which is bad for the reliability of the service.

References

Ashendorff, A., Principe, M., Seeley, A., LaDuca, J., Beckhardt, L., Faber Jr., W. and Mantus, J. (1997) "Watershed protection for New York City's supply," *Journal of the American Water Works Association*, 89 (3), March.

Bakker, K. (2004) *An Uncooperative Commodity: Privatising Water in England and Wales*, Oxford: Oxford University Press.

Barbier, J.M. (ed.) (2000) "Evolution des consommations d'eau," *TSM, Génie Urbain, Génie Rural*, 2.

Barraqué, B. (1998) "Europäisches Antwort auf John Briscoes Bewertung der Deutschen Wasserwirtschaft," *GWF Wasser-Abwasser*, 139 (6).

—— (2000) "Are hydro-dinosaurs sustainable? The case of the Rhone-to-Barcelona water transfer," in E. Vlachos and F.N. Correia, *Shared Water Systems and Trans-Boundary Issues, With Special Emphasis on the Iberian Peninsula*, Lisbon: Luso-American Foundation and Colorado State University.

—— (2001) "De l'eau dans le gaz à l'usine à gaz," proceedings of the conference *Hydrotop 2001*, Marseille 24–26 April, C-068.

Brouwer, F., Heinz, I. and Zabel, T. (eds) (2003) *Governance of Water-related Conflicts in Agriculture. New Directions in Agri-environmental and Water Policies in the EU*, Dordrecht: Kluwer Academic Publishers (Environment and policy series).

Cambon-Grau, S. (1999) *Baisse des facturations d'eau à Paris entre 1991 et 1997: analyse des causes sur un panel de gros consommateurs de Paris rive droite*, mimeo; Paris: SAGEP (July), p. 71.

Chesnutt, T.W. and Mitchell, D.L. (2000) *California's Emerging Water Market*, paper presented at the French *Académie de l'Eau.*

Dickinson, M.-A. (2000) "Water conservation in the United States: a decade of progress," in A. Estevan and V. Viñuales (eds) *La efficiencia del agua en las ciudades*, Bakeaz y Fundacion Ecologia y Desarrollo.

Euzen, A. (2002) "Utiliser l'eau du robinet, une question de confiance. Approche anthropologique des pratiques quotidiennes concernant l'eau du robinet dans l'espace

domestique à Paris," unpublished PhD dissertation, Paris: École nationale des ponts et chaussées.
Fougères, D. (2002) *Histoire de la mise en place d'un service urbain public: l'approvisionnement en eau à Montréal, 1796–1865*, PhD dissertation, Montreal: INRS-UCS.
Hughes, T.P. (1983) *Networks of Power: Electrification in Western Society (1880–1930)*, Baltimore, MD: Johns Hopkins University Press.
Lave, L. (1981) *The Strategy of Social Regulation*, Washington, DC: the Brookings Institution.
Llamas, R. (2001) *Las aguas subterraneas en España*, Fundación Marcelino Botín.
Maver, I. (2000) *Glasgow, Town and City Histories*, Edinburgh: Edinburgh University Press.
Moss, T. (2000) "Unearthing water flows, uncovering social relations: Introducing new waste water technologies in Berlin," *Journal of Urban Technology*, 7 (1).
Netzer, D., Schill, M. and Dunn, S. (2001) "Changing water and sewer finance, distributional impacts and effects on the viability of affordable housing," *Journal of the American Planning Association*, 67 (4).
Okun, D.A. (1996) "From cholera to cancer to cryptosporidiosis," *Journal of Environmental Engineering*, 122 (6): 453–8.
Okun, D.A., Craun, G., Edzwald, J., Gilbert, J. and Rose, J.B. (1997) "New York City: to filter or not to filter?" *Journal of the American Water Works Association*, 89 (3).
Tarr, J. (1996) *The Search for the Ultimate Sink, Urban Pollution in Historical Perspective*, Akron, OH: Akron University Press.
Van Humbeeck, P. (1998) "An assessment of the distributive effects of the wastewater charge and drinking-water tariffs reform on households in the Flanders Region in Belgium," Report of the SERV (Sociaal-economische raad van Vlandern), May.
Zérah, M.-H. (1997) "Inconstances de la distribution d'eau dans les villes du tiers-monde: le cas de Delhi," *Flux, cahiers scientifiques internationaux Réseaux et Territoires* 30, October–December: 5–15.

PART V
Networks as Institutions

CHAPTER TWELVE

Networks and the Subversion of Choice: An Institutionalist Manifesto

Gene I. Rochlin

Readers of this book will recall the meltdown in early 2001 of the once robust electrical system in California.[1] For most of those living in the Golden State, the consequences of deregulation were nearly inconceivable, and the origins of deregulation were at best opaque. For those more expert in the nature of the electrical network and its institutions, it is the reverse: it is the blind acceptance of the origins of institutional destruction that is nearly inconceivable, and the responsibility for the consequences that remains opaque.

Although California may remain an exceptional case of the chaotic deconstruction of electrical networks,[2] the effects on other systems such as the American telephone system (once the best in the world), British railroads, and the airlines are held by many to be similarly disruptive to the "average" user, imposing social and sometimes political costs that may well outweigh any of the presumptive economic benefits. In effect, the social fabric, woven of manifold social contracts and constructed to ensure the equitable distribution of social services to advanced industrial societies while controlling the power of those who own and operate the networks, has been rent in the name of efficiency. Politics has become subordinated to economics, collective interests to individual interests, and the common good to profit. To put it in its most compact metaphor, we have been "sandbagged."

What follows is not a professional paper in the usual sense nor is it a dispassionate critique, rather, it is something between a work-in-progress and a manifesto[3] – part analytic, part polemic, largely unfinished, and inadequately documented. I apologize for that last (but only for that last). Documentation for many of the social and political costs of the destruction (one might even say demolition) of historic large technical systems still remains fragmentary, anecdotal, and subject to the usual tiresome challenge that it will all work out if we only give it time and stop trying to interfere with the working of markets.[4]

One of the most striking characteristics of early twenty-first-century society is the broad acceptance of the belief that the "deregulation" and/or "privatization" of such essential infrastructure networks as electricity, telecommunications, financial services, rail services, and, perhaps, air traffic control and urban transit systems, can only be properly discussed in economic or, perhaps, techno-economic terms. Even to argue that such large technical systems are primarily socio-technical systems, that the services they provide have been designed to supply a variety of social and political services (many of them invisible) seems to be treated as a romantic idea, a holdover from the once-powerful ideals of socialism and social democracy that belonged to a previous century.

From professional journals to television reporting, there are few challenges to the claims that deregulation, decentralization, and computer-mediated management and integration are not only technically and economically superior to their previous hierarchically ordered and centralized forms, but that they "empower" users (or customers) by letting them socially shape the outcomes (La Porte 1996: 60–71). Modernity meant control, at the expense of efficiency. Post-modern forms were to promote both economically superior performance and socio-political liberation (Rochlin 1993; 1996: 55–59). The discursive essence of the Thatcherite–Reaganite claim to the political superiority of de-institutionalizing large technical systems, of converting them from hierarchies to markets, might, ironically, be put as: modern humans were born in chains, but will everywhere now be free. But the reality is radically different. Having broken what were claimed to be their visible chains, humans are led to deny the costs, and the more insidious means by which they increasingly become technically, economically, and socio-politically bound by the means and mechanisms of "free market" rules, structures, and coordination requirements. The costs of socialism and market-regulating social organization were (and still are) widely denounced. The social and informational costs to the average person of surviving amid neo-anarchic markets are almost completely ignored.[5]

None of this is meant to absolve the owners and operators of the large technical systems from the accusation that they have, at various times, been politically manipulative, unresponsive to challenges, insensitive to complaints, and slow to adapt to changing user/customer wants and needs. AT&T did, indeed, fight mightily against losing control even over installation of telephones not made by their wholly owned Western Electric subsidiary, and the irregularity and poor physical condition of British Rail before it was taken apart is, by now, legendary. And other examples abound (Kraus and Duerig 1988; "Britain Off the Rails" 2001). What is remarkable, however, is how easily these shortcomings and faults were used as a rationale to deconstruct the entire system, without at the same time pointing out that they were, in principle, subject to regulatory intervention to a degree that the (vastly under-stated) costs of the alternatives would not and could not be. Re-regulation, predicted by regulatory theory because some of the consequences are politically unpalatable, and because regulators are reluctant to leave things entirely to the market, meant that price ceilings and access controls remained in place (Harris and Milkis

1988). As a result, the network systems were not so much deregulated as deconstructed. As Peltzman and Winston point out in their recent (pro-deregulation) collection:

> When the United States began in the late 1970s to deregulate network industries – energy, transportation, and communications – the hard work appeared to be over. Once intractable political forces were overcome, and as deregulation moved forward it seemed only a matter of time before markets instead of regulators would fully determine the allocation of resources in these industries. More than twenty years later, markets are functioning in these network industries – and so are regulators.
>
> (Peltzman and Winston 2000: vii)[6]

But with a significant difference. The networks can be re-regulated, to some extent, but they cannot be coherently reassembled.

Private Goods, Public Bads

Theories of public administration have long understood that striking a balance between public and private organizations is as important for maintaining the balance of power in a democracy as balancing public and private interests (Moe 1991: 106–129). There are many services that might be best performed, or at least regulated, by public agencies rather than private organizations either to avoid undue concentration of power or to ensure that the public well-being is not sacrificed for economic gain (Simon 1995: 273–294; Simon *et al.* 1991). But the rush to switch network functions and regulations from the public to the private sector has proved to be an immense and uncontrolled social experiment, whose outcomes have been negative as well as positive, and whose prognosis remains mixed. As Herbert Simon stated in the 2000 John Gaus lecture of the American Political Science Association:

> The many experiments with privatization of services that had previously, for good or indifferent reasons, been supplied by public agencies, are beginning now to show us that switching to the market/business-organization system is not a sovereign remedy for all administrative ills.
>
> To illustrate what I have in mind, I need merely mention the complex mixture of gains and losses that deregulation of the air transportation industry has brought to its customers (in spite of rosy reports of fare savings). The same can be said of deregulation and privatization of energy distribution, education, and communications, all of which are faced today with perplexing economic and organizational problems. I could add other examples, notably the prison industry, which has not become a magical cure for criminal tendencies as a result of experiments in privatization.

> Nor can we say that we have solved all of the organizational problems posed by public goods and by such externalities as those associated with preservation of the environment. Experience has indicated that a wide range of essential services can be provided better by government than by any private business arrangement thus invented.
>
> (Simon 2000: 754).

But what are these "public goods?" The classic definition in economics is that such goods are jointly supplied, indivisible, and non-excludable. However, as Olson points out: "Students of public finance have [. . .] neglected the fact that *the achievement of any common goal or the satisfaction of any common interest means that a public or collective goal has been provided for that group*" (Olson 1971: 15, emphasis in original). Such goals or interests are not, however, limited to the quantity or tangibility of goods and benefits, but also to their "quality," which includes, *inter alia*, distributional equity, and ease of access. Moreover, those public goods that derive from large technical infrastructure networks also become elements of the social construction of public space, so that failure to provide them equitably, efficiently, and, in most cases, so effortlessly that they become almost invisible, has major effects on citizen perceptions of the quality of their lives and the nature of their society (Lewinsohn-Zamir 1998: 377–406).

Even if we limit ourselves only to those conditions that are included in the economic analysis of markets, the formal provision of full and complete information and open means for taking advantage of competition is a necessary condition for them to be fair and equitable. But it is not a sufficient condition unless that information and those means can be acquired and made use of by all potential consumers without imposing undue time or informational burdens on them. The evidence for the truth (or at least accuracy) of the oft-repeated economistic claims for the general superiority of deregulation and competition, even in large technical systems that provide non-substitutable goods or services is, therefore, disingenuous, since it ignores the remarkable increase in social and transaction costs that may be imposed on individual users. These come in four primary flavors:

- *direct costs* – the inability of many individual users to manage the information in such a way as to actually reap the benefits;
- *indirect costs* – time lost in conducting transactions;
- *the once-familiar but now apparently forgotten costs of the "knowledge burden"* – the direct and indirect costs of having to master a wealth of new information in order to function at all;
- for many of these systems, *the increased direct and indirect costs of acquiring the necessary information* for use, or maintenance, and, in many cases, the time needed to affect them.

This argument is strongly reminiscent of Ulrich Beck's theory of individualization. Beck (1992) argued that while "modernist" industrial society presented a highly structured risk environment to individuals, "reflexively modern" risk society removes them from historically proscribed social forms and commitments and reduces their security by de-legitimating traditional knowledge, institutions, and guiding norms. As one critic put it:

> Where to live, what to eat, where to take a vacation, what clothes to wear, with whom to mingle and to have sex with is [now] up to the individual. And it is not like in simple modernity any more, when the social democrats took care of the risks . . . the reflexive burden is placed upon the shoulders of the individual.
>
> (Almas 1999: 5)

Beck further points out that what at first appears as greater freedom – from class, family, and tradition – also entails a much greater range of pluralistic choices. But because there is far less room for not making choices, and the choices that must be made are more numerous, more demanding, and yet highly circumscribed, the individual actually ends up less free than before.

Moreover, while there is some evidence that even corporations and other powerful social entities and institutions best placed to reap returns are not making significant gains (as measured by the controversy over the rate of increase in productivity over the past decade), little study has been made of whether, and to what extent, individuals, particularly those individuals not fortunate or educated enough to belong to the growing upper-middle class of high-tech adepts, have made gains that outweigh the costs – including the costs of learning how to use the systems effectively.

In *Sorting Things Out: Classification and Its Consequence*s, Bowker and Star argue that:

> In the past 100 years, people in all lines of work have jointly constructed an incredible, interlocking set of categories, standards, and means for interoperating infrastructural technologies. We hardly know what we have built. No one is in control of infrastructure; no one has the power centrally to change it. To the extent that we live in, on, and around this new infrastructure, it helps to form the shape of our moral, scientific, and esthetic choices. Infrastructure is now the great inner space.
>
> (Bowker and Star 1999: 320)

Alas, it would appear that "someone" does indeed have the power to change it, albeit perhaps not centrally, and not consciously. As the most fundamental of the technical infrastructural systems become radically reconfigured (Summerton 1994), as not only their component parts but their rules, regulations, standards, and categories are deconstructed, so, in turn, is our social and political infrastructure (Osborn and Jackson 1988: 924–947). Services that were once accepted as not requiring conscious thought or deliberate choice now need to be attended to – in some cases not just once, but frequently and continuously.

And as some of the cases discussed below illustrate, the reconfiguration of that "great inner space" also alters our perceptions of the social world we live in, and our role(s) in it. The direct and, in principle, measurable social costs are, at least, acknowledged by some of the economic analysts of deregulation – albeit too often as "externalities" (Borenstein and Bushnell 2001: 46–52). There are other costs that are more difficult to quantify that arise from the deconstruction of both tangible and intangible infrastructure. Even the artifacts themselves, the instruments and instrumentalities of the networks, are deeply implicated in the ways in which the elements, processes, rules, standards, and narratives of large technical systems have become both agency and structure in our lives. But even less traditional economists usually regard only the humans as agents. They miss the importance of changes in the structure, operation, and regulation of the infrastructure because they neglect both the role of nonhumans as actors in the construction of our lives, and the role of artifacts, instruments, and the physical and technical environment in shaping our perceptions of our cognitive and social environments (Latour 1993, 1999; Hutchins 1995). The world in which we live is very different cognitively and politically as well as economically and socially from what it was when the telephone company provided horizontally and vertically integrated services, was the sole supplier of legendarily indestructible terminal equipment, did all the installation and maintenance, and took real pride in the quality of its service.

A Little History

As Thomas P. Hughes pointed out in his seminal work on electrification, and continued to explore in his subsequent work, the history of large technical systems revealed not so much the "diffusion" of technology as it did the self-conscious effort by inventors and entrepreneurs to create a market for their rail, power, and communications networks by creating a network of users (Hughes 1983, 1989). And, consciously or unconsciously, those users made a series of trade-offs and accommodations when they chose to enter into the networks. Shipping by rail did not allow the flexibility of choice in choosing a shipper as had traditional overland methods (Goddard 1994). Small independent electrical companies were gradually merged into large companies with more stable and extensive grids (Hughes 1983). The integration of the American telephone system replaced an ad hoc bricolage of small and independent companies with different rates and equipment (Fisher 1992).

The gradual integration of various infrastructure systems into networks was not, however, allowed to proceed on a purely *laissez-faire* basis. Governments seeking to stabilize the growth of macro-capitalism in the face of growing social movements struck a deal between citizens and capital, mediating the dual roles of the individual as consumer and citizen. As Graham and Marvin have pointed out for the case of cities, domestic access to such networked services as telephone, water, gas, and electricity became seen as a norm – as part of the social contract (Graham and Marvin 1994). Moreover, as transportation and communications spurred the drive for spatial and regional cohesion and, at least, a modicum of equity, government regulation became proactive with respect to rates and services, ensuring, for example, that new users of electrical grids were not charged the marginal costs of hookup, or, at least in principle, that railroad rates were set by distance rather than ease of access.[7]

As these systems grew into actual or de facto monopolies, various regulatory bodies or other means of governmental control were brought into being to protect users/customers from exploitation – in the US, for example, the Interstate Commerce Commission (ICC) for railroads, the Federal Communications Commission (FCC) for communication, and various public utility commissions to regulate electricity, telephones, and even water. As services became more extensive and more complex, and as populations, capabilities, and expectations grew, the structure of the networks themselves and the structures that controlled or regulated them coevolved into extensive, diverse, and highly bureaucratized complexes. These, in turn, increasingly began to demonstrate some of the well-known shortcomings and rigidities of bureaus – at times reaching proportions that only a Gogol could fully explicate.

Nevertheless, the period of the Cold War from 1948 to the 1980s partook of what John Ruggie (1983) has characterized (in the international context) as "the compromise of embedded liberalism." As Kate O'Neill recently put it: "This was also the era of the welfare state: many Western countries put in place extensive social security frameworks and national health services to protect their populations and provide what were seen at the time as basic social rights" (O'Neill 2001). In many Western countries, the provision of large technical system services such as the post, telegraph and telephone (integrated as PTT in some countries, spread among a mix of public and regulated private systems in others), electrical, and railroad systems (nationalized in some countries, heavily regulated private systems in others) were also embedded. But embedded liberalism was severely undermined by the rise of free-market ideologies in the West, while the collapse of socialism in the East reinforced the tendency to identify shortcomings and failures as the inevitable consequence of large integrated systems – whether government controlled or government regulated.

The international consequences of this transition, ranging from forced privatization through "structural adjustments" imposed by the IMF, the dismantling of the welfare state, and the opening up of free trade – leading, eventually, to the formation of the WTO (and, in turn, Seattle and other protests) – have been extensively addressed elsewhere. The internal consequences of the homologous internal transition have attracted less attention – except in those cases where system disaggregation has led to the export of production to poor countries with cheap and easily exported labor.

How, then, are we to understand the consequences of the dismantling and dismemberment (some would say "vivisection") of the large technical systems whose formation and growth were so central to the growth and development of twentieth-century technical societies? An early example, often overlooked, actually took place during the era of high modernity – the broad-scale destruction in the 1950s of American mass transit systems, first in favor of buses and the private car, and, later, to the private car over all (Goddard 1994). Widely touted as freeing Americans from the tyranny of transit schedules, the inflexibility of fixed infrastructures, and the forced company of strangers, it imposed upon them instead the tyranny of traffic schedules and regulations, the inflexibility of parking, and the forced company of other autos in traffic (Flink 1990). Similarly, the deregulation of other historically regulated networks in the US, such as electricity, telephone services, and airlines, was supposed to increase flexibility and provide a wealth of new options as well as lower costs (Peltzman and Winston 2000). It has done none of these. What it has done is add to the life of end users the need to master all kinds of new knowledge to use the systems efficiently.

Whether traditional or "infomated," modern or "post-modern," all socio-technical networks are subject to manipulation by those who are best placed to wield all forms of social power – political, economic, and cultural. These same actors not only support, but actively propagandize the wonders and benefits of the new networks, which are increasingly being subjected to both traditional monopolization through buyouts and control and new forms such as domination of infrastructure. In an era when regulation continues to be under attack and individuals continue to be seduced by the ideology that public institutions are, at best, inefficient and, at worst, evil and the only relevant levels of politics are the individual and the state, there would seem to be little that can be done. I argue here that the history of networks within the framework of this idea is one that merges radical, anti-central-government ideology (sometimes referred to as the Thatcher/Reagan revolution), academic free-market ideologies, and the ability of entrepreneurial capitalists to forge sufficient political alliances to deconstruct stable network systems purely for the sake of financial gain. I also note, with some pain, that in many cases this hodgepodge of greed, ideology, and narrow self-interest has been aided and abetted by many from the "left" who have similarly demonized both central governments and large, "natural monopoly" industries, albeit usually for quite different reasons.

Privatization and Piracy

Deregulation actually began before the Thatcherite revolution and the transformation of *laissez-faire* ideals into anti-government ideologies. In the US, awkward and clumsy rate regulation had become a joke by the 1960s, handcuffing the railroads while at the same time severely restricting competition from the trucking industry (Harris and Milkis 1988). With the construction of an elaborate interstate highway system, the pressure to open up trucking and remove it from the control of the Interstate Commerce Commission was overwhelming. With railroad profits in sharp decline, American railroads, too, were "partially" deregulated in the 1970s (Goddard 1994). But the deconstruction of the system was not deliberate, as it was to be later in the UK (Guy *et al.* 1999). Indeed, there was some hope that deregulation would strengthen them financially. As Grimm and Winston state: "The railroad industry is perhaps the only US industry that has been, or ever will be, deregulated because of its poor performance under regulation" (Grimm and Winston 2000: 41).

But it was the triumph of the new conservatism, nurtured by Thatcher in the UK and carried to the US under the flag of Reagan, that led to an all-out attack on all aspects of government interference with markets. Even the most traditional forms of regulation were to be considered evils – and not necessary ones, either. In the UK, or other countries with national utilities, national airlines, and national railroads, that meant attacks on the government's ability to run a business. In the US, where heavily regulated private industry played the same role, it meant mobilizing ideological tools under the guise of antitrust, antimonopoly and rationalizing economic ones on the grounds that the heavily regulated public industries were not economically efficient.

Among the first to promote and then exploit the new wave of ideological and economic arguments for deconstruction of large technical systems under the guise of "deregulation" and/or "privatization" were entrepreneurial firms seeking to profit by entry into systems from which they had been excluded by law and regulation geared to the theory of natural monopolies. These entrepreneurial firms ranged from small regional airlines to companies such as MCI (who were instrumental in bringing about telephone deregulation) to large banks and utilities seeking to diversify by engaging in a variety of economic activities from which they had been excluded. (This ranged from AT&T's interest in the computer business to PG&E's seeking to make profits in unregulated activities such as investing in the booming California real estate market.)

In a variety of circumstances, the proliferation of competitive actors was rapid and highly visible. Regional airlines sprang up all across America. MCI, Sprint, and others competed actively for long-distance phone customers. AT&T did, indeed, go into computers. PG&E separated out most of its power generation and sold it to an unregulated subsidiary, while retaining the transmission and distribution system. Independent operators moved in rapidly to run trains over the British rail system, while the "natural monopoly" core, the rails themselves, and the signaling system, remained in the hands of Railtrack. And, at first, and in some places, prices fell and service increased. Or, at least, prices fell for some users, and certain highly visible aspects of service, generally defined, increased. But what were the hidden and systemic costs?

Three Easy Piecemeals

There is far too much literature on the deregulation of large networked systems to review in this short chapter, and there are many issues that cannot be addressed here – such as deregulation of banks and financial institutions, transnational and global networks, and global communications. What follow are three "snapshots" organized and reviewed specifically to point out the range and scope of unintended social consequences. The public perceives the impacts of these on their lives. The comparative indifference of politicians, economists, and corporate actors to those perceptions does much to destabilize confidence in the role and reliability of the networked infrastructure and the institutions that manage it.

The Kahn Game

Airline deregulation in the US was not the result of a spontaneous outcry from either the public or from government officials. If there was any political impetus at all, it came from business. But the one person most closely identified with airline deregulation was economist Alfred Kahn, who almost single-handedly pushed it through, defended it, and was actually more than once heard to blame both the airlines and the users for the subsequent problems, arguing that they did not behave the way they were supposed to. As Dempsey and Goetz put it:

> Before deregulation, the United States enjoyed what was universally acclaimed to be the "world's finest system of [air] transportation." Our service was excellent, our fleet was young and technically efficient, labor enjoyed stability of working conditions and decent wages, and inflation-adjusted airfares had been falling for four decades. But Alfred Kahn thought he could do better.
>
> (Dempsey and Goetz 1992: 335)

Since the book was written, the fleet has improved, but labor conditions have definitely not. And the economic benefits, measured against theoretical models of what the fares "would have been" under regulation, are still argued about (Dempsey and Goetz 1992: 243ff.).

Kahn had been active in airline regulatory reform in the 1970s, which resulted in breaking informal price agreements and lowering fares, and thought that if reform was good, deregulation, by increasing competition, would be even better. What he did not foresee was that there were, indeed, significant economies of scale in airline operations, that there were social barriers to entry that outweighed the low economic ones, and that market power could easily be exercised by the major airlines. Kahn assumed that the industry could not become concentrated and that competition would provide the service to smaller cities with less frequent flights that regulation had imposed on the airlines as part of the price of entry into lucrative, heavily traveled markets. In essence, he both overestimated the degree to which market power could counter cherry-picking and the extent of marginalization in small markets that would occur once the competition had been suppressed: I clearly remember the day in the early 1980s when my connection from New York City to Ithaca, New York took place in a four-seat, single-engine Beechcraft.

There are more frequent flights, and considerable savings – at least for business travelers. But according to Dempsey and Goetz, the airline industry had, between 1978 and 1992, lost more money than the accumulated profits since the birth of air travel. There were hundreds of bankruptcies, an increase in price discrimination, a bewildering proliferation of fares, and a loss of public confidence in the quality and reliability of service. The growth in hub-and-spoke operations, held up by economists as one of the major benefits of deregulation, often means longer total flight times and a greater possibility of delays and missed connections. Planes are arguably more crowded, and less comfortable, and the decline in in-flight service needs no further mention. The fare structure has moved from baroque to bizarre – on a single flight, passengers sitting in equivalent seats may have paid dozens of different fares. The cost of flights is governed by competition and traffic, not distance. Most small airlines that sprang up in the wake of deregulation have gradually been squeezed out of the market or absorbed by bigger airlines, and even these whales are beginning to consume each other. Only one carrier (America West) remains of the 58 that started up between 1978 and 1990. New ones do continue to start up, but their prognosis is mixed. What may change things is the advent of essentially "no-frills" (i.e., no service) airlines that can compete effectively on price. This is how Southwest became a success competing against the majors. (How safe, or reliable, these prove to be remains an open question, since they work their aircraft very hard. Moreover, when even one aircraft is down for service or with a problem, the ripple effects can paralyze the entire operation.)

For a while, service was increasing and complaints were dropping. But as the majors moved in and took over the market, "public exasperation over deteriorating airline service [was] fueling no fewer than five bills in Congress for a so-called bill of rights for the nation's airline passengers" (Armstrong 2001: B1). The coalition pushing for the legislation includes the American Society of Travel Agents as well as Consumers Union and the ad hoc National Airline Passengers Coalition. Complaints include frequent flight delays, overbooking, cancellations, lost luggage, and seemingly interminable waits at airports. According to Armstrong's article, 25 percent of all flights in the US in 2000 were delayed, and the on-time figure of 72.6 percent was the worst on record. One of the major features in the proposed legislation is to allow passengers to sue in state courts under state consumer protection laws – a right that was specifically banned in the original Deregulation Act of 1978. Moreover, and in contrast with the expectation that business travelers would be the major beneficiaries of deregulation, business people's complaints about the tiresome indirect flights of hub-and-spoke and the growth in missed connections are driving forces behind the push for new legislation.

Whether or not airline deregulation actually resulted in reduced airfares in some highly traveled markets, it certainly raised them in other less heavily used markets, and some communities have vastly reduced (as well as more expensive) access. Ticket pricing has become a nightmare of complexity, and independent travel agents are increasingly marginalized and may eventually disappear altogether. Free to compete, but forbidden to collude, large airlines simply coordinate. They also reduced service, the size of seats, and the amount of footspace. Doing this allowed them to squeeze their smaller competitors out of the market entirely. And although growth in flight frequency is said to be a major contributor to the growing problems, that growth was a specific goal of deregulation – and yet, as with rail travel in the UK, telephone service in the US (see below), and transmission and distribution of electricity under deregulation, none of the entities created by the deregulation acts was fully responsible, or adequately funded, even to do network maintenance, let alone to fund and coordinate the necessary growth in the system's infrastructure.

Phone-y Arguments

When the Federal Communications Commission (FCC) was created in 1934, the first item on its agenda was an inquiry into the Bell Telephone system. Even at that early date, the FCC saw problems in AT&T's giving exclusive rights for equipment hookups to its wholly owned subsidiary, Western Electric. The FCC's pursuit of Bell resumed in 1949, eventually resulting in a negotiated settlement in which AT&T was allowed to keep Western Electric, and Bell Telephone Laboratories, under very strict conditions: Western Electric could only sell to AT&T; AT&T would confine itself to carrier operations; and technical information developed at the jewel in the company's crown, the Bell Laboratories, would become available to anyone who applied for it.[8]

By the late 1960s, there were rumblings of discontent again. The computer industry was just beginning its period of rapid growth, and AT&T was challenging the limits put on it. Perhaps more to the point, in 1968 AT&T lost an important case on the hookup of non-Western Electric equipment (the Carterphone decision), and its monopoly looked very vulnerable to outsiders seeking to use some of the new technology to move into the telecommunications field (as it now became known). In 1969, the FCC authorized an independent common carrier, MCI, to construct microwave radio systems that would compete with AT&T. This was the wedge that would eventually split AT&T apart (La Porte 1996: 60–71). Although MCI claimed it had no intention of competing with AT&T's long distance service, it did, in fact, do so almost immediately. The FCC ruled that this was illegal, but was, in turn, overruled by a Federal court in 1977. MCI was quickly followed by Sprint and others when it became clear that they would enjoy what amounted to a *de facto* government subsidy by being exempt from paying the local operating companies and independents for their local service hookups – which AT&T had to continue to do (Kraus and Duerig 1988: 89).

In the meanwhile, the case of the *United States v. AT&T* was proceeding at a snail's pace through the US courts. Circumstances did not favor AT&T. The case finally went to trial just after Ronald Reagan's inauguration, essentially preventing an out-of-court settlement. And it went by the luck of the draw to a judge who leaned toward enforcement of antitrust laws rather than to one who saw AT&T as a quasi-public service organization. Perhaps Walter Annenberg, as quoted in Kraus and Duerig, said it best in 1983:

> The breakup of America's telephone system, acknowledged to be the most efficient in the world, will affect nearly every aspect of our society. How this came to pass is a frightening example . . . of what can happen here to a company recognized as one of our major national and defense assets. It is a company that grew by its own efforts and with its own resources to become the world's largest business, whose Bell Laboratories led us into the information age.
>
> (Kraus and Duerig 1988: 11)

The best telephone system in the world, running the best private research laboratory in existence, was peremptorily disassembled, its laboratory spun off to seek profit, its local companies to compete when and wherever they could, and its long-distance arm to compete with the entrepreneurs of MCI, Sprint, and the others who had fought to bring it down. Ironically, AT&T, freed to compete, turned out finally to be perhaps the largest (and, eventually, perhaps even the fiercest) tiger in the telecommunications jungle, and might not only outlast but absorb its weaker competitors in every sector. But, of course, it will do so with, at best, minimal regulation and across a broader spectrum of telecommunications infrastructure. And Lucent, which inherited the newly privatized Bell Laboratories and several other of the more "public interest" functions, suffered from a series of poor financial decisions that drove it to bankruptcy. (It has been resurrected as a company named Avaya.)

In the meanwhile, some of the more pronounced social consequences include not only the burden placed on consumers (and discussed elsewhere in this essay), but the constant telephone harassment by companies seeking to get you to switch your service provider, "slamming" (illegal switching of service without informed consent by the end user), and the turmoil over the constant reassignment of area codes as more and more independent players (particularly cell-phone services) grab off large blocks of numbers within existing area codes.

Before legislation was passed allowing subscribers to block most unwanted sales calls, many of us frequently exploded in anger at yet another telephone call from MCI, Sprint, or even AT&T, pushing us to change our phone service over to them. We can testify that the transition was not without some unmeasured individual social cost. But, of course, there have been other costs as well. In California, at least, the imposition of fixed charges for long distance access meant that phone bills actually increased for users with a low volume of long distance calling. The remaining vestige of regulation is regularly lobbied to do away with "inefficient cross-subsidies," by which is meant demolishing what others would describe as the fundamental social contract of ensuring that all users, everywhere in the system, have equal access at equal costs regardless of their income or location.

Power Plays

The deregulation of electrical systems seems to have also originated more with economic theorists and ideologues of market-based competition than with any experience in the real world. While I cannot speak with confidence about the origins of deregulation in the UK (which was prior to, and often invoked as an argument for, deregulation in the US), I note that even Paul Joskow, a supporter of deregulation, has written:

> Several supply-side indicators show, however, that the electric power sector in the United States has performed fairly well over time. In particular, it has supplied electricity with high levels of reliability, investment in new capacity has been readily financed to keep up with (or often exceed) demand growth, system losses . . . are as low or lower than those in other developed countries, and electricity is available almost universally. The traditional system was efficient and reliable in dispatching generation plants; making cost-reducing short-term energy trades between generating utilities; maintaining network reliability; and dealing with congestion, unplanned outages, and system emergencies . . . .
>
> Average real electricity prices in the United States fell rapidly from the early 1900s until the early 1970s. Indeed, during this time period the US electric power sector had one of the highest rates of productivity of any major industry in the US economy.
>
> (Joskow 2000: 119)

So, what happened? In the early 1980s, with deregulation ideology in full swing, rates began to rise, fueled by higher interest rates, increased demand, poor investments in overly large and costly generation facilities (e.g., nuclear power plants), and a mandate under the Public Utility Regulatory Policy Act of 1978 (PURPA) to pay top dollar to smaller "alternate" (i.e., environmentally friendly) power plants. Although internally many of the electric utilities took considerable pride in the quality and reliability of their service, they were increasingly criticized from all sides (Hirsh 1999: 504). Even though rates fell again from the mid-1980s into the 1990s, business firms in areas where rates remained comparatively high (e.g., San Francisco and New York) complained vigorously. The utilities also received considerable "consumer" criticism from ratepayer organizations as well as from a variety of environmental, NIMBY, and antinuclear movements.

In the midst of this deregulatory turmoil, in the midst of a booming economy, many of the utilities saw an opportunity to rid themselves of the burdens of historical regulation, which included a ban on investments outside the electrical system as well as public utility commission rate regulation (Borenstein and Bushnell 2001: 46–52). In California (and in New York), free-market ideologues, economic theorists of deregulation, utility critics, and the utilities themselves actually combined forces to put deregulation into place

(Hirsh 1999: 504). But those charged with reconfiguring the system faced a set of problems. Electricity networks consist of several "subsystems" – generation, transmission (over long distances), distribution (to the end user), and such "soft" coordination services as billing, power trading, load management, and efficiency promotion. In both California and New York the solution was to set up an "independent system operator" (ISO) to manage the actual physical network, and separate out generation for competition. In California, the electrical utilities, freed of responsibility for maintaining the transmission system, quickly divested themselves of generation as well. This created a market system for power, complete with bids and auctions, that is far too complex (and varied) to discuss here.

What happened in California, however, was the direct consequence of a failure by the political parties to understand that electricity is not a "commodity." Not only is it essential, and non-substitutable, its marginal price is not set by what you are willing to pay to have it, but how much you are willing to pay to avoid *not* having it.[9] Worse yet, from a consumer's point of view, the amount of power available at any given time is finite. These two facts combined mean that the supply curve does not increase smoothly with price; instead, the price rockets upward when the supply limit is approached. Was there ever a greater incentive for generators to exercise market power and hold their supply off the market? In the winter of 2000–1, Californians found themselves paying truly outrageous prices for marginal power – many hundreds of dollars per megawatt-hour. But the end-user rates had been frozen as part of the same deregulation deal (with some objections from consumer advocates who firmly believed that costs would fall). And as many observers point out, the California crisis was the result of supply marketing, not of energy supply itself (Williams 2001: 626–30).

To try to rescue an increasingly strident social and political situation, Governor Davis moved to make the State of California a buyer of electricity, seeking thereby to stabilize the market. What the state had forgotten, however, was that regulated public utilities make open transactions. Only if the utility commission retains its integrity will the public readily accept the legitimacy of the outcome (even if they criticize the details). By making private deals for undisclosed sums, the governor and the state agencies exposed themselves to furious criticism. (To give a single example, the state purchased advance power contracts at moderately high prices as insurance against another surge – a standard business practice. When prices actually fell sharply because of unexpectedly low demand, the state was attacked for having "wasted" taxpayers' money.) Moreover, with the state becoming an owner of the grid as well as taking on major responsibilities for negotiating power contracts, there is a certain irony to the term "privatization." In 2001, California enjoyed an unusually cool summer and other generators came on line, thus avoiding a repetition of the crisis. But the state went into a budget crisis, and the once-popular governor not only saw his approval ratings plummet, but was recalled and replaced in a special election.

Once upon a time, the users of electricity paid almost no attention to the elaborate network that supplied them. It was a standing joke in our program at Berkeley that as far as most end-users were concerned, the electricity that came from the wall plate might as well have been generated by mice in the walls. Power failures were exceptional, horrid events, and the cost of electricity was, for most users, a minor part of household budgets. All of that has changed, and explicit social costs, including the breaking of an implicit social contract, are yet to be tallied.

Summary

Yes, indeed, it turns out that unregulated (or insufficiently regulated) free markets for non-substitutable services are hostile environments for small independent firms. Apparently, most of the entrepreneurial promoters of gaining market access forgot that a return to the open competitive markets favored the kind of predatory behavior that characterized the early phases of industrial capitalism. There was no way to level the playing field as if all parties to the newly deregulated activities had been created equal. While the UK (and others) struggle with the preferential gate and airport access given to established airlines, the consolidation of major US airlines has grown to the point where the government is seriously considering stepping in again. There is a real prospect that Virgin will soon be the only major rail service provider in the UK. AT&T seems to be re-emerging as the dominant player in telephone provision, threatening to marginalize even once-robust competitors such as MCI. And large banks are eating even equally large ones in a frenzy of takeovers that has, for the first time in the US, created a situation where the owners, and top management, of neighborhood banks and lending institutions may be located in other states far away from regional concerns, sometimes thousands of "social" as well as geographic miles away.

Does any of this really matter? I would argue that it does. In many cases the breakup of large technical systems under the guise of deregulation did not lead as promised to the emergence of effective competition and competitive markets. Instead, those with the greatest or most effectively used market power are moving to re-aggregate the system, but this time largely free of the regulatory and government controls that restrained them from exploiting either their customers or their workers. Moreover, the search for efficiency and profit has in many cases led to the conversion of these systems from social institutions to pure monopolies. Where such companies as PG&E or AT&T once saw themselves as guarantors and providers of essential social services as well as machines for generating shareholder profits, they now seem to focus almost entirely on the latter. And that, too, has social costs that have been largely unexplored.

Working Together

Taylorism, the scientific management system of Frederick Winslow Taylor, first entered the universe of large technical systems through its intersection with

"Fordism" and the productive mode of the assembly line as eventually manifested in that early triumph of the large technical manufacturing system, the River Rouge Plant (Hughes 1989). Sadly, little of the extant literature on large technical network systems directly addresses the effects of system growth and integration on the various labor forces that were enrolled. Systems as described tend to be made up of machinery, plants, communications systems, entrepreneurs, social and political "environments," and government regulations. Nevertheless, the Taylorist–Fordist ideology and approach were not confined to industries that mass-produced "hardware" such as automobiles or washing machines, with their simple structures and standardized outputs, or to offices that performed a narrow range of simple tasks. The growing sophistication of process at the end of the nineteenth century had resulted in a major internal reorganization of industries whose output was now based largely on scientific and engineering knowledge.

For many of the large technical systems, the parts were rearranged into functional specialties, each of which increasingly made use of a limited subset of the necessary labor skills. In the telephone industry, for example, there was one division made up almost entirely of operators, another of repair personnel, another for accounting, and so on. Similarly, electrical systems were roughly divided into lines and distribution, power generation, system management, and so on, with finer divisions within each category. As the networks grew, they tended to reduce the uncertainty of their productive as well as external environments by incorporating critical tasks (and, therefore, the labor, skills, and specializations associated with them) as much as was reasonably possible. But such mega-systems as AT&T or large power companies were able to move past the blindness of Fordism and instill in their workers a sense of pride in the company's accomplishments – sometimes, and not erroneously, compared with the *esprit de corps* of elite military organizations.

Over the past several decades, there has been an ongoing debate about the nature and role of labor in an increasingly post-Fordist society – it is a well-populated subdiscipline. My point is that, on reflection, the focus and trend of this continuing debate and its relationship to large technical (network) systems can be framed as part of the grand narrative of modernity, in which progressive growth of systems and hierarchies, and the ever more seamless integration of technology and labor were seen as part of the inevitable course of techno-history. From this master narrative was constructed the never-ending story of independent, autonomous, skilled workers drawn (a classic Marxist would say forced) into the maw of the mega-machine where they became alienated labor – stripped first of autonomy, then of skill, and then turned into "human productive elements." Because this was (and still largely is) the dominant analytic paradigm, what is often overlooked is that it addresses structures and relations within a productive system dealing with a labor force that is always already present.

Most of the historical literature also ignores the effects on the lives of workers outside the context of the plant, or on the lives of other workers in the heterogeneous networks whose work depends upon it. In *American Genesis*, Hughes extends the analysis of the famous case of Frederick Taylor and the "Dutchman" (Henry Noll) to go beyond the question of how Noll was manipulated, or how his productive capacity increased. As Hughes puts it:

> Taylor singled out a "little Pennsylvania Dutchman who had been observed to trot back home for a mile or so after his work in the evening about as fresh as he was when he came down to work in the morning." After work he was building a little house for himself on a small plot of land he had "succeeded" in buying. . . . [Noll] moved the forty-seven tons of pig that the Taylorites had decided should be the norm, instead of the former twelve and a half tons, and soon all the gang was moving the same and receiving sixty percent more pay than other workmen around them. We are not told whether [Noll] was still able to trot home and work on his house.
>
> (Hughes 1989: 194–195)

With the socialization reforms of management theory in the 1930s and thereafter, abuses such as these were controlled, but only for workers within the system. But here I must return to anecdotes from personal experience, since the literature seems almost silent.

My brother-in-law, Roger, worked in Detroit, as part of the large technical system that was Ford Motors in its heyday – when the spirit of the River Rouge plant still lived, and Ford tried to internalize almost every aspect of production, from smelting iron to making its own paint. But like many thousands of others in the "auto industry," Roger had a specialized craft skill that was tied not to production per se, but to the annual model changeover, and was, therefore, handled by contract rather than internalization. Roger was a draftsman and spent many long hours drafting the drawings necessary to convert, for example, an idea for a new brake into something for which the production machinery could be adapted. He was also responsible for seeing to it that the new brake "fit" into the complex whole that would be the finished vehicle. Working outside the system, and paid on what amounted to a piecework basis, he worked long hours, when he worked, often had no work, and did not receive the benefits of health plans, insurance, and retirement that Ford employees enjoyed.[10]

A similar experience occurred as part of our group study of the operation and maintenance of nuclear power plants (Rochlin and von Meier 1994). The operator of a plant we recently studied had a large labor force devoted to maintenance as well as to operations – in this case a much larger one than is typical in the industry (Rochlin and von Meier 1994). Major tasks such as turbine overhaul and maintenance, are contracted out to the original equipment manufacturer, or other highly specialized firms. But during a major maintenance shutdown, we found that some of the workers involved were not permanent employees either of the operating company, the plant, or the major subcontractors. They were "nuclear gypsies," a mobile labor force that moved from plant to plant doing seasonal outage work. Although much more highly skilled than migrant farm labor, and far more highly paid, many of the conditions of work – unpredictability, seasonal variation, enforced mobility, lack of health plans, retirement, or insurance – were very similar. Moreover, under the pressure of deregulation, the plant we studied has shut down many of the divisions that it formerly kept on its own payroll (such as computing) and shifted to outside contracting on a competitive basis instead. And that is increasingly true for the electrical system as a whole, as it is fragmented by deregulation, and as its component pieces shift to a competitive market structure instead of a regulated one.

In one industry after another, the decomposition of the large technical systems has altered the terms and stability of employment in ways that can affect us, the users and customers of the system, as well as the workers themselves. The effects on the systems as their parts are pulled apart are, in some cases, glaringly obvious, as in the case of separating electrical generation from transmission from distribution, or train operators from those who own and maintain the tracks. Other effects due to the paring off of system support services are more subtle. As many of us have noted, the telephone centers for reaching technical support, or customer services, are often not only subcontracted out but located far away where labor costs are lower and unions nonexistent. And often that imposes a non-trivial burden on us, the end users, as well.

Conclusion

So, where do we go from here? If present conditions are any gauge, the large technical systems that so characterized the twentieth century will continue to be chopped up, piece by piece, in the name of an anti-socialist, free-market ideology disguised in socio-political terms as autonomy, empowerment, and choice, and in economic terms as market competition and efficiency. The economic benefits will continue to accrue to the largest and most powerful actors in the system, while endless (and nearly costless) symbolic rewards are dispensed to the average person.

Does this mean that large networked technical systems (LTS) will disappear? Of course not. Although agency may change radically, structure will not, and in many cases cannot. Railroads, ripe for revival even in the US, will still require rails and signals, scheduling and coordination, stations, station agents, and reservation systems – and, if these are no longer managed by the same entity, external agents for coordination, integration, and, yes, regulation. Air travel similarly still requires airports, route and terminal scheduling, air traffic control, reservation and information systems, and, yes, always, air safety regulation. Electrical supply still takes place over networks of power, elaborated, interconnected transmission and distribution networks, and still requires exquisitely elaborate real-time monitoring to balance supply against load. Even telecommunications will still take place largely over wires rather than wireless transmission, and, in any case, still requires standardization of signal, rights of access, and (often forgotten) some entity to ensure that each terminal has one and only one unique access code (a.k.a. phone number), and that this valuable resource is allocated reasonably (if not always fairly).

These systems were dismantled and privatized for the sake of presumptive economic efficiency and equally presumptive competition. But if present trends continue without intervention, a few of the more powerful actors, particularly those built of the larger fragments of the former systems, will begin to aggregate specific functions until they have reorganized into large, near-monopoly, system-scale firms, but this time relieved of the burden of providing social services and more independent of government regulation and control.

For some, that may simply mean becoming providers of special services within the network (such as having a near monopoly on electrical generation) while the network infrastructure itself is maintained and operated by a private, public, or mixed private–public entity whose resources and financial condition are inadequate to the task of network maintenance and repair, let alone expansion and modernization (such as Railtrack in the UK, or whatever replaces the ISO in California). For some, such as AT&T, the larger airlines (in the US), the dominant rail carriers (in the UK), or the electrical utilities in several countries, it may mean regaining something close to their a priori power, without regulation or requirement to provide service equity. For still others, operating under the aegis of international liberalism, EU barrier demolition, and the WTO, it could mean assembling a network that is more supranational than transnational, operating in a global system whose governance mechanisms are designed almost entirely to foster open market liberalism, and international arrangements that prevent, or at least hamper, the ability of national governments to assert some degree of regulatory control, or at least some modicum of oversight. In contrast to the balancing forces of environmental movements that prevented the race to the bottom in environmental regulation, there are still very few voices, except those in the streets, willing or able to take on the least-common-denominator approach embedded in the NAFTA and WTO regimes (Vogel 1995; Bhagwati 2001: 15–29).

Where will it all end? That question is more than rhetorical. It has already been suggested that various governments could completely divest themselves of postal services, of air traffic control, of urban transport systems, and of highway maintenance. It has even been suggested, with at least a trace of humor, that the last remaining large technical system operating along socialist lines will be the military. But, hey! Why not privatize that as well? Do you need someone to intervene in a dispute in Central Africa? To stabilize borders in the Balkans? To stamp out drugs in Central America? Why put up with the expensive infrastructure and long-term costs of a professional military when you can simply put out a Request for Proposal and get the far more efficient private sector to respond? (Ironically, some military and security functions have, indeed, been contracted out in Iraq.)

On what may or may not turn out to be a more serious vein, the real issue with future large technical systems may lie in domains that are not simple extrapolations of the iron-and-fixed-infrastructure systems that are the core of the historical literature. Some of the systems that are the most fundamental to the emergence and stability of a society that is based more on the exchange of bits and information than the trade in words and goods are already far beyond the scope of governmental control, at any level (Rochlin 1997; United States, President's Commission on Critical Infrastructure Protection 1997).

For those of us who had hoped that the manifest effects of deregulation would lead to a back reaction, there is scant hope. There are some signs of resistance in Europe, perhaps the most notable being the very strong negative reaction in London to moves to privatize the tube. Similarly negative reactions to the privatization of air traffic control are not as well grounded politically as, for example, movements against genetically modified foods that are grounded in the manifest issues of immediate risk and safety ("A Better Way to Run a Railway" 2001: 55). However, the election of George W. Bush and the dominance of both the Executive branch and both houses of Congress by the Republican Party have given aid and comfort to the corporate lobbyists who would further dismantle government regulations and consumer protection in the name of private profit. As Robert Scheer so eloquently put it in an Op-Ed piece in the *Los Angeles Times*:

> Capitalism is falling apart. Tires explode, utility rates skyrocket, pharmaceuticals kill patients, telephone service is a mess, airports are gridlocked, broadcasters rip off scarce airwave spectrum for free, and salmon in the Northwest are becoming transgendered and unable to breed. Even successful dot-commers are an endangered species. Yes, Virginia, we do need government regulation. Not to build socialism but to save capitalism, because the market mechanism left to its own devices inevitably spirals out of control.
>
> (Scheer 2000)

Nor are these effects confined to traditional infrastructure networks such as electricity and railroads whose regulation was grounded in natural monopoly theory. The radio frequency spectrum is now up for grabs to the highest bidder, as are the last pristine forests and the Alaskan Wildlife Refuge. The Financial Services Modernization Act of 1999, allowing the merger of banks, insurance companies, and stock brokerages, threatens to create unregulated networks that control not only financial services but credit records and other personal information. Biotechnology companies are merging with pharmaceutical firms to create immensely powerful international corporate networks to promote new drugs, genetically modified organisms, and genetically modified foods.

As the hardware firms of the nineteenth century evolved into tightly woven networks at the turn of the twentieth century, the Progressive movement in America arose in response to an enormous public outcry over the economic and political abuses that followed upon the unrestrained control of essential goods and services. For most of the century, its ideals of social justice and distributional equity, although frequently mangled, managed to survive a constant tattoo of ideological attack (Hofstadter 1986:185). It is, therefore, doubly ironic that the growth of electronic and computer networks, which made possible the most extensive and accessible social networking in human history, was also accompanied by the ideological triumph of efficiency over equity, of neo-classic free-market individualism over social contractualism. We are increasingly the servants of the networks that were ostensibly created to serve us. And governments, once seen as the protectors of individuals against corporate interests, are now criticized by left and right alike as unwieldy, controlling, and inefficient – except when they use their powers to free those interests from social control.[11]

In her marvelous review of the debate over consumer preferences v. citizen preferences in the rationalization of government intervention to provide public goods, Lewinsohn-Zamir argues for the nefarious effects of spillover: "Yet, it is doubtful that the preference ranking of people who are basically self-interested in their private lives will undergo radical transformation in their public lives" (Lewinsohn-Zamir 1998: 389). What we are now observing, at least in American society, would appear to be reverse spillover. When self-interest becomes the dominant ideology of interaction in the public sphere, how could it do other than reflexively spill back into their private ones? And in rough analogy with the arguments of Beck and Giddens about risk societies, the mutual reinforcement socially deconstructs confidence in public as well as private institutions – including those that design and operate the networks and those that regulate them. And if one is looking for evidence of this last point, one has to look no further than the newspapers in my home state of California.

Afterthoughts

What, then, is next? Even the most classic texts on regulation and deregulation point out that every wave of deregulation tends to lead to a third step – re-regulation – as citizens try to regain some control over the perceived negative effects of complete deregulation (Harris and Milkis 1988). But these networked systems, once taken apart, are irreversibly reconfigured into post-modern, or at least "post-Fordist" forms that can no longer be fully controlled by political bodies trying to impose minimum restraints on markets and competition. In California, at least, the attempts at re-regulation have possibly made a bad situation worse and a fragile situation unstable, and perhaps guaranteed that we will be doomed forever to live in a twilight zone between traditional regulated utilities and genuine free markets. Nor, apparently, is the situation any better in New York, or in the UK (Wolfram 2000: 48; Borenstein and Bushnell 2001: 46–52; Berenson 2001). So, I can offer the reader a choice of endings, depending on whether one believes that the future lies with more privatization and less regulation:

> Things fall apart; the center cannot hold;
> Mere anarchy is loosed upon the world,
> The blood-dimmed tide is loosed, and everywhere
> The ceremony of innocence is drowned;
> The best lack all conviction, while the worst
> Are full of passionate intensity.
>
> (W.B. Yeats, *The Second Coming*)

Or, alternatively, whether one believes that we are faced with a future of fragmented attempts to re-regulate the deconstructed networks:

> Hegel remarks somewhere that all facts and personages of great importance in world history occur, as it were, twice. He forgot to add: the first time as tragedy, the second as farce.
>
> (Karl Marx, *The Eighteenth Brumaire of Louis Bonaparte*)

Notes

1 I apologize for the essentially Anglo-Saxon focus of the paper. There are two reasons for this. The first, and perhaps most persuasive, is that arguments such as these are addressed largely to the more developed countries, and that the US (and, to a lesser extent, the UK) have been (and still are) among the least socialist (or at least the least social-democratic) of all of the countries of the OECD. The two most glaringly obvious facets of this in the US are that the term "socialist" is an accusation, not a description, in American politics, and, empirically, that the dogged and ideological defense of the costly, badly distributed, and almost wholly privatized medical systems in the US is made in the face of all evidence of the social (and medical) superiority of the European approach. The second reason is more reflexive; since so little has been published on the subject in professional journals, I have had to rely to an unusual degree on newspaper stories, television coverage, anecdotal evidence, and personal discussions to construct incomplete narratives of life under deregulation.

2 Although there are some suggestions that California's trials and tribulations are precursors of events to be replicated elsewhere in the US, and possibly abroad, as other electrical systems are similarly deregulated, some of the aspects of California's plan, including preventing the utilities from signing long-term contracts with generators, will hopefully not be repeated.

3 And a rather moderate manifesto at that. For a more forceful (and articulate) critique of deregulation from the "old" political left; see, for example, Scheer (2000).

4 The economic literature on deregulation is truly enormous, and far too extensive even to be catalogued here. A useful recent survey is the collection edited by Peltzman and Winston (2000), with articles that praise the economic benefits of airline, railroad, telephone, and electrical utility deregulation. Except for the penultimate article on electrical utility deregulation by Paul Joskow, these uniformly pay at best lip service to indirect social costs.

5 Indeed, this is a common general critique of neo-classical economics as a tool, and as a discipline. In the past few years there have been many critiques, and one entire field (ecological economics) that have grown up as a counter. One example of particular interest within the framework of this chapter is the work of Addelson (1995) whose

challenge to neo-classical economic theory focuses on the relationship between peoples' understanding, their social worlds, and the decisions they make. I thank Paul Baer for bringing this to my attention.

6 Remarkably, this otherwise splendid collection omits discussion of the banking and financial industry, perhaps making the frequent error of failing to see networks without the presence of dedicated physical infrastructure. Equally regrettably, the subject is also too vast to cover in this brief essay.

7 In the US, at least, the manipulation of railroad rates by the Interstate Commerce Commission led to a rate book that was somewhere between baroque and outrageous. This became one of the spurs for deregulation of transport. See, e.g., Harris and Milkis (1988).

8 This last was by no means trivial. The Bell Laboratories soon became as preeminent in basic science as in applied. Among the inventions that emerged from the Laboratories in the next few years were the transistor and the laser, both of which ultimately had effects on the world almost as far-reaching as the original telephone.

9 I am indebted to Dr Katie Coughlin of the Lawrence Berkeley National Laboratory for this observation.

10 The kind of work he did is now obsolete, and Roger's successors are masters of CAD/CAM software rather than ink and paper; but the firms still contract out.

11 The most glaring example, internationally, is the WTO regime, which is widely criticized as a means for subordinating government interests in protecting workers, or the environment, to the "requirements" of free trade (McMichael 2000: 466–74; O'Neill 2001).

References

Addelson, M. (1995) *Equilibrium vs. Understanding*, London and New York: Routledge.

Almas, R. (1999) "Food Trust, Ethics, and Safety in Risk Society," *Sociological Research*. Online. Available at: http://www.socresonline.org.uk (accessed February 2001).

Armstrong, D. (2001) "Late on Arrival," *San Francisco Chronicle*, 30 March: B1, B4.

Beck, U. (1992) *Risk Society*, London: Sage.

Berenson, A. (2001) "New York Faces Prospect of Its Own Energy Troubles," *The New York Times*, 25 February. Online. Available at: http://www.nytimes.com/2001/02/25nyregion/25POWE.html (accessed 26 February 2001).

Bhagwati, J. (2001) "After Seattle: Free Trade and the WTO," *International Affairs*, 77: 15–29.

Borenstein, S. and Bushnell, J. (2001) "Electricity Restructuring: Deregulation or Regulation," *Regulation*, 23: 46–52.

Bowker, G.C. and Star, S.L. (1999) *Sorting Things Out: Classification and Its Consequences*, Cambridge, MA: MIT Press.

Dempsey, P.S. and Goetz, A.R. (1992) *Airline Deregulation and Laissez-Faire Mythology*, Westport, CT: Quorum Books.

Economist, The (2001a) "A Better Way to Run a Railway," 17 March: 55.

Economist, The (2001b) "Britain Off the Rails," 17 March: 18.

Fisher, C.S. (1992) *America Calling: A Social History of the Telephone to 1940*, Berkeley and Los Angeles, CA: The University of California Press.

Flink, J.J. (1990) *The Automobile Age*, Cambridge, MA: MIT Press.

Goddard, S.B. (1994) *Getting There: The Epic Struggle Between Road and Rail in the American Century*, Chicago, IL: University of Chicago Press.

Graham, S. and Marvin, S. (1994) "More Than Ducts and Wires: Post-Fordism, Cities and Utility Networks," in P. Healey, S. Cameron, S. Davoudi, S. Graham and A. Madani-Pour (eds) *Managing Cities: The New Urban Context*, London and New York: John Wiley.

Grimm, R. and Winston, C. (2000) "Competition in the Deregulated Railroad Industry: Sources, Effects, and Policy Issues," in S. Peltzman and C. Winston (eds) *Deregulated Industries: What's Next?*, Washington, DC: Brookings Institution Press.

Guy, S., Graham, S., and Marvin, S. (1999) "Splintering Networks: The Social, Spatial, and Environmental Implications of the Privatization and Liberalization of Utilities in Britain," in O. Coutard (ed.) *The Governance of Large Technical Systems*, London and New York: Routledge.

Harris, R.A. and Milkis, S.M. (1988) *The Politics of Regulatory Change*, New York: Oxford University Press.

Hirsh, R.F. (1999) *Power Loss: The Origins of Deregulation and Restructuring in the American Electric Utility System*, Cambridge, MA: MIT Press.

Hofstadter, R. (1986) *The Progressive Movement, 1900–1915*, New York: Simon & Schuster.

Hughes, T.P. (1983) *Networks of Power: Electrification in Western Society 1880–1930*, Baltimore, MD: The Johns Hopkins University Press.

Hughes, T.P. (1989) *American Genesis: A History of the American Genius for Invention*, New York: Viking Penguin.

Hutchins, E. (1995) *Cognition in the Wild*, Cambridge, MA: MIT Press.

Joskow, P.L. (2000) "Deregulation and Regulatory Reform in the US Electric Power Sector," in S. Peltzman and C. Winston (eds) *Deregulated Industries: What's Next?*, Washington, DC: Brookings Institution Press.

Kraus, C.R. and Duerig, A.W. (1988) *The Rape of Ma Bell: The Criminal Wrecking of the Best Telephone System in the World*, Secaucus, NJ: Lyle Stuart, Inc.

La Porte, T.R. (1996) "High Reliability Organizations: Unlikely, Demanding, and at Risk," *Journal of Contingencies and Crisis Management*, 4: 60–71.

Latour, B. (1993) *We Have Never Been Modern*, Cambridge, MA: Harvard University Press.

Latour, B. (1999) *Pandora's Hope: Essays on the Reality of Science Studies*, Cambridge, MA: Harvard University Press.

Lewinsohn-Zamir, D. (1998) "Consumer Preferences, Citizen Preferences, and the Provision of Public Goods," *Yale Law Journal*, 108: 377–406.

McMichael, P. (2000) "Sleepless Since Seattle: What Is the WTO About?," *Review of International Political Economy*, 7: 466–74.

Moe, T.M. (1991) "Politics and the Theory of Organization," *Journal of Law, Economics, & Organization*, 7: 106–29.

Olson, M. (1971) *The Logic of Collective Action: Public Goods and the Theory of Groups*, Cambridge, MA: Harvard University Press.

O'Neill, K. (2001) "Capitalism, Cappuccinos, and Chaos: Transnational Protest and the Bretton Woods Institutions," Chicago: International Studies Association Annual Meeting, March.

Osborn, R.N. and Jackson, D.H. (1988) "Leaders, Riverboat Gamblers, or Purposeful Unintended Consequences in the Management of Complex, Dangerous Technologies," *Academy of Management Journal*, 31: 924–47.

Peltzman, S. and Winston, C. (eds) (2000) *Deregulated Industries: What's Next?*, Washington, DC: Brookings Institution Press.

Rochlin, G.I. (1993) "Defining High-Reliability Organizations in Practice: A Taxonomic Prolegomenon," in K.H. Roberts (ed.) *New Challenges to Understanding Organizations*, New York: Macmillan.

Rochlin, G.I. (1996) "Reliable Organizations: Present Research and Future Directions," *Journal of Contingencies and Crisis Management*, 4: 55–9.

Rochlin, G.I. (1997) *Trapped in the Net: The Unanticipated Consequences of Computerization*, Princeton, NJ: Princeton University Press.

Rochlin, G.I. and von Meier, A. (1994) "Nuclear Power Operations: A Cross-Cultural Perspective," *Annual Review of Energy and the Environment*, 19: 153–87.

Ruggie, J.G. (1983) "International Regimes, Transactions and Change: Embedded Liberalism in the Postwar Economic Order," in S.D. Krasner (ed.) *International Regimes*, Ithaca. NY: Cornell University Press.

Scheer, R. (2000) "These Messes Are What Deregulation Gets Us," *Los Angeles Times*, 26 December. Online. Available at: http://www.robertscheer.com/1_natcolumn/00_columns/122600.htm (accessed 1 January 2001).

Simon, H.A. (1995) "Organizations and Markets," *Journal of Public Administration Research and Theory*, 5: 273–94.
Simon, H.A. (2000) "Public Administration in Today's World of Organizations and Markets," *PS-Political Science & Politics*, 33: 749–56.
Simon, H.A., Smithburg, D.W., and Thompson, A.V. (1991) *Public Administration*, New Brunswick, NJ: Transaction Publishers.
Summerton, J. (ed.) (1994) *Changing Large Technical Systems*, Boulder, CO: Westview Press.
United States, President's Commission on Critical Infrastructure Protection (1997) *Critical Foundations: Protecting America's Infrastructures: The Report of the President's Commission on Critical Infrastructure Protection*, Washington, DC: GPO.
Vogel, D. (1995) *Trading Up: Consumer and Environmental Regulation in a Global Economy*, Cambridge, MA: Harvard University Press.
Williams, J.C. (2001) "Strictly Business: Notes on Deregulating Electricity," *Technology and Culture*, 42: 626–30.
Wolfram, C.D. (2000) "Electricity Markets: Should the Rest of the World Adopt the United Kingdom's Reforms?," *Regulation*, 22: 48.

AFTERWORD

After Words

Seymour J. Mandelbaum

These brief comments at the end of this anthology are not intended as a conclusion offering to the overwhelmed reader a convenient interpretation of what the 13 chapters mean when they are taken all together. Nor have I tried to provide a summary of each chapter in turn. That has already been done in the excellent introduction written by the editors. I certainly have not attempted to place myself in the role of critic, leaving that vital but often unpleasant task to independent reviewers.

This afterword is informed by a modest objective. I've attempted to provide a bridge between the carefully crafted sentences of the authors and the less-disciplined practices of readers who will control those sculpted texts by investing them both with meanings and with practical implications.

I read the edited text of the anthology in April 2004. I had, however, read earlier versions of the chapters in April 2001 when they were presented at an international seminar on *The Social Sustainability of Technological Networks.* When I opened the package of revised seminar essays, I recognized (as most readers would not) that the title, *Sustaining Urban Networks: The Social Diffusion of Large Technical Systems*, was new.

The contrast of the two titles intrigues me. I don't, however, mean to suggest that there is anything terribly portentous about the decision to amend the title. There isn't. I suspect that if I had asked the editors to explain the reasons for the shift they would have provided a simple and unremarkable justification. I have not, however, risked losing my deconstructive play before the sober authority of the editorial committee: I haven't asked for an explanation.

It seems to me that each of the titles points to a slightly different bridge between the formal essays and the discursive practices of readers. Taken together, they reveal a productive ambiguity surrounding the hard images of matters technical and technological; of the representation of networks and of sustainability.

The ordinary and familiar conception of "sustainability" is grounded in an image of Nature as a bundle of materials that the clever minds and contriving hands of human beings convert into the "natural resources" that our artifices require. The gifts of nature are not fixed. We are adept at creating new resources from old materials: electricity from rapidly flowing water and tiny logic machines from sand.

The pride we take in our skill does not free us from anxiety and the specter of hubris. Is it possible that we will exhaust Nature's gifts? that our new creations will cease to delight us? that we have stolen the patrimony of our own children and will be damned for our profligacy? that we will be endlessly frustrated by the failures of what we imagined would be an endlessly triumphant science?

Those questions (and others like them) dominate public discussions of sustainability and will shape the ways in which some readers will read this anthology and, I suspect, inevitably misread it. This volume is not an account of resource exhaustion, species extinction, global warming or the Brundtland Commission. The title of the seminar points, instead, to the social issues associated with the functioning (or dysfunctioning) of technological networks. The form of the argument in some chapters is similar to John Rawls' insistence that polities cannot be stable if they are not just. In parallel terms some of the participants in the 2001 seminar struggled with the fear that the networks that carry water, information and energy over vast distances will be assaulted if they splinter the provision of vital utilities, destroying the rough equality that maintains the life-sustaining flows.

The title applied to the anthology sharpens this social apprehension. The networks are vast but the vulnerable points of attack and the threats of collapse – no light, no water, no news, no electricity – are urban. Prudently attending to the ultimate constraints of a site or resource is relatively easy compared with the constant need *to sustain* large networks against both deliberate and neglectful destruction. (These words and every reading of the anthology will come after 9/11 even in the chapters that do not march to that terrible beat.)

The new title, however, should not be understood as providing a remedial strategy after the colon. The idea that networks would be protected if they were continuously *diffused* across continents, if all local inequalities were temporary and all misapprehensions corrected by intimate knowledge carries with it great hopes and profound skepticism. Admittedly, the phrase after the title colon is meant as descriptive, rather than normative. Yet, the diverse approaches taken by the authors as they address the "digital divide," power black-outs, and dry water taps will shape the ways in which readers invest in this volume.

Author Index

Subject Index